PRACTICAL APPLICATIONS of COMPUTATIONAL INTELLIGENCE for ADAPTIVE CONTROL

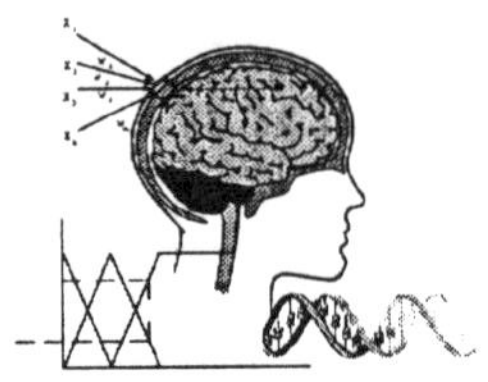

The CRC Press

International Series on Computational Intelligence

Series Editor
L.C. Jain, Ph.D.

L.C. Jain, R.P. Johnson, Y. Takefuji, and L.A. Zadeh
Knowledge-Based Intelligent Techniques in Industry

L.C. Jain and C.W. de Silva
Intelligent Adaptive Control: Industrial Applications

L.C. Jain and N.M. Martin
Fusion of Neural Networks, Fuzzy Systems, and Genetic Algorithms: Industrial Applications

H.N. Teodorescu, A. Kandel, and L.C. Jain
Fuzzy and Neuro-Fuzzy Systems in Medicine

C.L. Karr and L.M. Freeman
Industrial Applications of Genetic Algorithms

L.C. Jain and Beatrice Lazzerini
Knowledge-Based Intelligent Techniques in Character Recognition

L.C. Jain and V. Vemuri
Industrial Applications of Neural Networks

H.N. Teodorescu, A. Kandel, and L.C. Jain
Soft Computing Techniques in Human-Related Science

Beatrice Lazzerini, D. Dumitrescu, A. Dumitrescu, and L.C. Jain
Evolutionary Computing and Applications

Beatrice Lazzerini, D. Dumitrescu, and L.C. Jain
Fuzzy Sets and Their Application to Clustering and Training

L.C. Jain, Ugur Halici, Isao Hayashi, S.B. Lee, and S. Tsutsui
Intelligent Biometric Techniques in Fingerprint and Face Recognition: Practical Applications

Zhengxin Chen
Computational Intelligence for Decision Support

L.C. Jain
Evolution of Engineering and Information System and Their Applications

H.N. Teodorescu and Abraham Kandel
Dynamic Fuzzy Systems and Chaos Applications

Larry Medsker and L.C. Jain
Recurrent Neural Networks: Design and Applications

PRACTICAL APPLICATIONS of COMPUTATIONAL INTELLIGENCE for ADAPTIVE CONTROL

Charles L. Karr

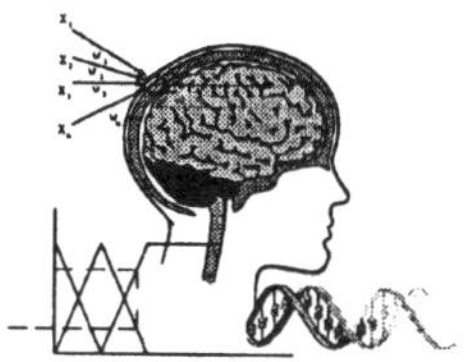

CRC Press
Boca Raton London New York Washington, D.C.

Acquiring Editor: Dawn Mesa
Project Editor: Sylvia Wood
Marketing Managers: Jane Stark / Barbara Glunn
Cover design: Dawn Boyd
Manufacturing Manager: Carol Slatter

Library of Congress Cataloging-in-Publication Data

Karr, C.L. (Charles L.)
Practical applications of computational intelligence for adaptive control / by Charles L. Karr
p. cm. -- (International series on computational intelligence)
Includes bibliographical references and index.
ISBN 0-8493-2069-0 (alk. paper)
1. Adaptive control systems. 2. Computational intelligence.
I. Title. II. Series
TJ217.K37 1999
629.8′36—dc21 99-10490
CIP

International Standard Book Number 0-8493-2069-0
Library of Congress Card Number 99–10490
Printed in the United States of America 1 2 3 4 5 6 7 8 9 0
Printed on acid-free paper

Dedication

During my life there have been so many people who have contributed to making me the man that I am. However, there is one person in particular who, like the Lord, has always been behind me, supporting me with all that she is. Oftentimes, because of my personal shortcomings, I have neglected to give her the acclaim and attention she deserves. I know that this does not make up for all the times that I have slighted her, neglected to thank her, and disappointed her; no matter,
I dedicate this book to my mother, Betty Karr. I love you.

TABLE OF CONTENTS

Foreword

Dr. Donald A. Stanley was one of my mentors, my business associate, and my close friend. We began this book project together and had hoped to complete it together. Unfortunately, Don succumbed to an illness and died before we could complete the project.

Near the end, Don Stanley graciously asked to have his name removed from the project; he felt strongly that the work was basically mine, and that I should be the sole author. That is the way Don was – unselfish, unassuming, and always striving to do what is fair. Out of respect for Don, I have agreed to his request. And I thought strongly about re-writing the Preface of the book since the Preface was co-authored. However, I believe this would be a mistake. Thus, I have left the Preface unchanged; it appears as Don and I originally wrote it together.

This book is about research conducted at the U.S. Bureau of Mines during my and Don's tenure there. The research described is not the work of one or even two individuals, rather it was accomplished through the effort, drive, and insightful nature of numerous people – one of whom was Donald A. Stanley. It is my hope that the reader benefits from the efforts of our research group, including the mistakes we made along the way. Don Stanley was a big part of our success.

Donald A. Stanley was extremely bright. When he set his mind to something he was like a bulldog – he just would not let it go until "we got it right." I owe a lot of my success to the lessons he taught me. I miss my good friend Don dearly.

Charles L. Karr
The University of Alabama

Preface

This book is about intelligent, adaptive process control. It brings together ideas from the field of *computational intelligence*, a part of the larger field of artificial intelligence, including fuzzy mathematics, genetic algorithms, and neural networks, and uses these ideas to develop a generic architecture for accomplishing adaptive process control. In the development of this architecture, the requisite tools are described and then demonstrated on a number of problems. Moreover, most of the examples provided are of interest in industrial settings (although some simple examples are provided in the beginning so that the reader can focus on technique and not be overburdened with the complexity of the problems being solved).

The focus of the book is on practical applications. It is meant to provide practicing engineers and scientists with the information they need to solve control problems in industry and in academia. Worldwide research is providing a vast body of literature on each of our tools, but most of this information is not needed for solving process control problems. In fact, it is amazing what a small portion of fuzzy logic, genetic algorithm, and neural network theory is required to develop simple, yet extremely effective, intelligent, adaptive systems. If the reader is interested in solving difficult control problems, then this book is a good place to start.

The intended audience for this book is the practicing professional who has a technical background. However, do not let the term "technical background" intimidate you. We do not presuppose an understanding of computational intelligence, control systems, or even calculus. Since our intent is to reach a broad spectrum of people, we present examples from a number of scientific fields. To successfully reach such a diverse audience, the tone of the book is intentionally casual; terseness is almost always sacrificed in the interest of clarity.

Now that we have a rough indication of what this book is, let's discuss for a moment what it is not. We feel it is important to begin by building a relationship based on trust with the reader. Thus, if this book is not for you, let us save you time and expense: don't buy it.

WHAT THIS BOOK IS NOT

This book is not a text book on process control theory. There are already many good texts on control theory — linear, nonlinear, and adaptive. Besides, such textbooks require far more high-level mathematics than we need for our purposes. Complex numbers, Nyquist plots, convergence theorems — we cringe at the thought of the time it takes to gain an appreciation (much less an understanding) of these concepts. Besides, we have to wonder: are these concepts really necessary for developing effective, comprehensible software and for attaining excellent control of even the most complex physical systems? Our experience suggests that they are not.

This book is not a textbook on the theoretical analysis of fuzzy logic, nor of genetic algorithms, nor of neural networks. There are a number of excellent books on each of these three techniques, and they are used in college courses across the world. Books such as these provide great detail on many aspects of the various techniques. In fact, in our humble opinion, they cover many more aspects of the techniques than you really need. Although such theoretical texts are certainly necessary, this is not one of them.

This book is not a collection of works by various experts in the field. In fact, there are no experts involved, only us. Edited volumes in which numerous famous folks write about their earthshaking theoretical research efforts are great to have on one's bookshelf. However, when you have a problem that must be solved and solved quickly, these books are usually the last reference you want to reach for. Such books are often quite thought provoking and useful on that infamous "higher plain," but they are rarely consulted when the successful solution rubber meets the difficult real-world problem road.

This book is not a collection of case studies in which complex solution techniques are applied to toy problems. There is nothing more frustrating than reading a journal paper in which difficult-to-understand algorithms are applied to trivial problems. More often than not, when the algorithms are extended to problems of any consequence, they fail. Thus, this book presents only simple solution techniques that have been used to solve problems that are complex and typical of problems faced every day in industry.

A BOOK BASED ON EXPERIENCE

This book is based on a seven-year project using computational intelligence techniques for solving mineral processing problems, conducted at the U.S. Bureau of Mines, Tuscaloosa Research Center. It is an approximate chronological recap of the research project. The authors initiated the project, directed the research effort, and were the key players in the successes (and failures) experienced. Because of the Bureau's interest in industrial cooperation and in technology transfer, we were able to solve a number of difficult problems in and out of the minerals industry. For instance, this book contains example applications from the fields of aerospace engineering, chemical engineering, geology, and mineral processing; areas that superficially are not of direct interest to the Bureau of Mines, but areas that provided a forum for us to discover and test the ideas we were developing for the mineral processing industry.

The authors have very different backgrounds that actually served as a plus over the life of the project. Don Stanley was trained as a mineralogist. He worked for the Bureau of Mines for over 30 years. During his career he supervised projects in mineral surface physics, ceramics, and fine particle technology. He has worked with countless professionals from government, academia, and industry. Thus, Dr. Stanley brought some traditional ideas to the project. Chuck Karr, on the other hand, had never worked in the field of mineral processing when the project began. His education was in the fields of computational intelligence and fluid dynamics; he knew little of the traditional approaches to process control commonly used in industry. Thus, Dr. Karr brought some very nontraditional ideas to the project. Although this strange mix made for interesting and sometimes heated discussions over the years, it was not all bad. For instance, because neither of us had worked with process control problems, we had no preconceived notions of how process control should be done. Thus, we were forced to forge our own way in this area.

This book is, in a sense, a history of the research effort conducted at the Bureau of Mines. When the project began, non-adaptive expert systems, including fuzzy expert systems, were available to solve a number of difficult problems. However, we wondered at the time if such expert systems were all there was to computational intelligence. Asking this simple question began a seven-year journey that has been enlightening, exciting, satisfying, and often confusing. We discovered that the answer to the question, in a word, is no; there is far more to the field of computational intelligence than non-adaptive expert systems.

On this journey, in which we went from a prototypical non-adaptive expert system for phosphate flotation to an architecture for adaptive, intelligent control that has been applied to chemical processing, to a column flotation system, and to helicopter flight—we made some interesting discoveries, ran into some brick walls, and had a lot of fun. If you have aspirations for applying computational intelligence techniques to real-world problems, you will undoubtedly face some tough questions. We hope to help you address the issues you will face by relating to you some of the insights we gained and some of the technology we developed along our path.

The choices we made along the way were not the only ones possible at the time. However, they were, we feel, acceptable because they worked. We do not feel that the path we took to developing adaptive, intelligent control systems is the only one, or even necessarily the best one. We do feel that sharing our results has some potential benefits. Some will read what we did and immediately come to the same conclusions. Others will read what we did and immediately decide that there are better ways. Hopefully, the second class of readers will reciprocate our effort and tell us about techniques that can produce better control systems.

STRUCTURE OF THE BOOK

The book is divided into four parts, and these parts basically follow the order in which we implemented various techniques along the way.

Part 1 consists of four chapters dedicated to building fuzzy controllers. There is no detailed description of high-level fuzzy operators nor is there an attempt to prove convergence theorems. There is, however, a brief overview of why we chose fuzzy rule-based systems as our main tool for developing control systems. There is also a step-by-step procedure for building fuzzy systems, and this procedure is applied to three different problems of varying complexity.

Part 2 consists of five chapters on genetic algorithms. After a gentle introduction to genetic algorithms, these search algorithms, based on the mechanics of biologic evolution, are used to re-design the control systems presented in Part 1. Also, the importance of developing computer models is introduced. Computer models are needed in the realm of process control, especially in adaptive control, because they allow for rapid, inexpensive testing of alternative control strategies and for the training of adaptive control systems. In fact, the idea of modeling is so important to the systems presented in this book, there is an entire section devoted to it.

Part 3 focuses on developing computer models that can be used in adaptive control systems. This section presents three important topics: (1) genetic algorithms for tuning empirical computer models, (2) neural networks for process modeling, and (3) fuzzy mathematics for process modeling. Each chapter contains examples from industry.

Part 4 ties everything together. Adaptive, intelligent control systems are not problem specific. We will demonstrate their generality by solving problems from a number of different domains. The problems addressed include pH manipulation, chemical processing, column flotation, and helicopter flight control. Additionally, a problem from chaotic dynamics is solved. Since these systems can be large-scale, nonlinear, and coupled, they are frequently difficult to control. Yet, in each example the ideas presented earlier in the book permit the development of effective control systems. Additionally, each application chapter has a little different twist that is specific to the problem being addressed. Each problem has been selected to address an issue we faced over the life of the project.

ACKNOWLEDGMENTS

There is no question that this section will be incomplete; there are so many people who have made contributions to this book. Perhaps we should simply follow the format of the rest of this book and go somewhat chronologically. Thus, we start with the man who first introduced us to the field of computational intelligence, David E. Goldberg. Dr. Goldberg was Chuck Karr's dissertation advisor at The University of Alabama, and arguably the leading figure in the field of genetic algorithms. His course provided Chuck with an excellent overview of genetic algorithms. Dr. Goldberg patiently (and sometimes painfully) guided and prodded him through various applications of genetic algorithms, and taught him how to pay attention to the details necessary to conduct quality research. For this, and much more, Chuck is in this gentleman's debt.

Next, we owe a great deal to many people at the U.S. Bureau of Mines. Our supervisor, Bernard Scheiner, was infinitely patient when we would go off course and start working on "hair-brained" ideas. To his credit, when it was time to quit wasting time, he told us. Furthermore, on those few occasions when we were right and he was wrong, he admitted that "even a blind hog occasionally finds an acorn." Most importantly, this gentleman has served as a tremendous resource. He has over 30 years experience in the minerals industry and knows virtually everyone who is anyone in the field. Thanks Bernie; thanks a bunch.

Additionally, we need to thank two Bureau of Mines employees from the Washington, DC office: Bob Kaplan and Fred Schottman. Bob was head of the Minerals and Materials Science Division in the Bureau. Although it often appeared that we had led the research group "way out in left field," Bob stuck with us. Fred was head of the program under which this work was done. He had to answer to Bob, and often bought us valuable time.

So much for the people above us, now for the people on the front line. There is no question that thanks need to go to our co-workers and former and current students. We will not try to kid you: these are the folks who have done all of the work, we pretty much just "wrote it up." Since this book represents their contribution, we will simply try to list their names: Sandra Bodle, Charlie Burroughs, Barry Cole, Brod Davis, Cindy Dunn, Jen-Ho Fang, John Fleming, Mike Freeman, Ed Gentry, Tony Harper, Jeffrey Harris, Cy Jordan, Willie Larry, Don Meredith, Pat Moore, Kevin Mosley, Chad Phillips, Sandeep Sharma, Patricia Vann, Barry Weck, Mike Wrenn, and Dorian Yeager. We sure hope we did not leave anyone out, but we probably did.

In conclusion, let us just say that we, and we alone, are responsible for any errors or oversights in this book. We are sure there are mistakes, and for that we are truly sorry. In summation, the people we mentioned above are responsible for what is right with this book, while we contributed all that is wrong.

PART I
FUZZY SYSTEMS

Chapter 1

Why *Fuzzy* Process Control?

1.1 INTRODUCTION

Fuzzy logic can serve as a bridge between mathematics and language. Humans readily and efficiently manipulate quantities, concepts, and abstract linguistic terms. Computers, on the other hand, manipulate mathematical expressions with lightning speed. In fuzzy systems, traditional production rules are combined with fuzzy membership functions to produce systems that are easily comprehended by humans and effectively applied by computers. The fuzzy sets provide meaning to the linguistic terms used in the rules. There is nothing fuzzy about the logic, rather the sets used to define the linguistic terms are fuzzy.

Fuzzy systems are becoming increasingly popular as tools for solving difficult problems, and many products employing fuzzy logic have recently been marketed. Numerous types of problems have been solved effectively in a wide range of application domains. Process control is no exception. Fuzzy logic is being embraced by the controls community because it greatly simplifies the development of efficient and comprehensible systems.

When this research effort began, the standard for industrial process control was the proportional-integral-derivative (PID) controller. The setpoints for individual PID control loops were determined by human operators based on their experience with some particular system; different operators developed their own impressions of how the system should be controlled and adjusted the setpoints accordingly. The authors have witnessed situations in which an operator spends his or her entire eight-hour shift tweaking setpoints to get the system "running smoothly." The next operator on duty immediately changes these setpoints. When questioned as to why he changed the setpoints the previous operator spent an entire shift establishing, the operator generally responds with a more or less polite version of, "He's an idiot." The point is that there are differences from operator to operator in selecting setpoints. These differences are often so apparent that simple pattern recognition investigations of plant operating data identify when shift changes occur.

Because of the limitations of PID control, scientists and engineers in a wide variety of industries began to investigate alternative control strategies. When the current research effort began, researchers in the mineral processing industry were beginning to investigate artificial intelligence techniques for increasing the efficiency and improving the consistency of process control systems. In one research area they were beginning to investigate the use of expert systems for the control of mineral processing systems (Melema, et al, 1987).

Expert systems are rule-based systems that utilize the if-then production rules used to capture human decision-making abilities. These computer based systems have been used successfully in a number of problem domains (Tanimoto, 1987) including medical diagnosis, mineral exploration, and chemical compound analysis. The success they had achieved made them potential tools for attaining consistent, efficient process control. However, as will be pointed out in this chapter, these expert systems have shortcomings that make them poor vehicles for precise control.

This chapter will provide insight into why expert systems were not used as the basis for the control systems developed by the Bureau of Mines' research effort into adaptive process control. In keeping with the flavor of the book, this current chapter does not provide theoretical evidence, nor does it present convergence proofs regarding conventional control algorithms. Rather, it presents the qualitative arguments used by the authors in directing the research effort. Later in the book, fuzzy expert systems are introduced as extensions of traditional expert systems, and a link between fuzzy models and fuzzy control systems is forged. Finally, the modeling ability of a traditional expert system is compared with that of a fuzzy expert system where each is used to model a small data set.

1.2 A BIT OF HISTORY

Complex industrial systems are controlled acceptably every day by human operators. Generally, these operators have no background in complex mathematical control strategies. Instead, they gain a "feel" for the system's response to various control actions, and this "feel" allows them to manipulate the system effectively. One of the authors got a first-hand look at this method early in his career while working as a consultant for an oil and gas pipeline company. Consider the following account of his experience:

> *I was working for an oil and gas pipeline company in Atlanta, Georgia developing a computer model of transient waves in a gas pipeline pumping station. One day I wandered into a control room where an operator was overseeing an elaborate computer system that provided him with the information he needed to make control decisions. I began talking with him to find out just what his responsibilities were. I soon found that he was responsible for controlling all of the pumps and compressors on an oil and gas pipeline that ran from Houston, Texas up to somewhere in New Jersey. Needless to say, I was extremely impressed*

with this gentleman's ability to manipulate such a complex system. Surely, I thought, he used some kind of computer program to help insure the safe and efficient operation of this pipeline. I was taken aback to learn that his education consisted of "finishing most of the tenth grade." Being young and naive, I asked him if there wasn't a computer program somewhere he could use to make his job easier. He told me that some professor from a local university had tried to develop one, but it used some high level math instead of good old common sense and experience. The computer program was a failure and was used only by the professor in his classroom.

Here was an operator with less than a high school education who was doing a job that a university engineering professor failed to do with computers and mathematical methods. This was an observation that had a major impact on the direction of this research.

When this project was initiated, the mineral processing industry presented an even more challenging setting than the pipeline industry. This was true for several reasons including: (1) the minerals industry, in general, was not highly computerized, (2) the minerals industry did not have adequate sensors for process control, (3) the ore input to the mineral processing systems varied a great deal over time, and (4) the mathematical models available were inadequate or unproven. Thus, the minerals industry provided a tremendous opportunity for developing innovative process control systems.

The "textbook" approach to developing a control system begins by developing a computer model (a transfer function) for the system to be controlled. Then, the transfer function is manipulated using various mathematical techniques. However, the authors had neither the time, the money, nor the inclination to develop and validate the transfer functions for complex mineral processing systems. Besides, are such transfer functions really necessary when human operators can achieve a reasonable level of control without them? In our investigation of this question, we discovered the writings of a wise man who had already been over this road.

Lofti Zadeh wrote the first paper on fuzzy set theory in 1965 (Zadeh, 1965). This paper introduced a system that allowed linguistic concepts to be manipulated by computers. It provided a mechanism for incorporating linguistic terms so familiar to human operators into computer programs. His arguments for adopting such an approach were centered around two basic ideas. First, as systems become more complex, the possibility of efficiently controlling them using traditional mathematical techniques is reduced. This occurs for numerous reasons including the failure to accurately model the complex systems over all-inclusive ranges of operations and because complex situations must be modeled in bits and pieces. Second, there are numerous situations in which a high price is paid for unneeded accuracy. The high price paid for accuracy is apparent in many day-to-day activities such as the task of parallel parking a car; there is no need to be precise, just get the car in the space without bumping into one of the neighboring cars. In a similar fashion the human process control operator is effective because he just gets the process controlled.

Zadeh's early work was quite general. It did not focus on the use of fuzzy mathematics for process control, rather on its use in approximate reasoning tasks (Zadeh, 1968; 1971a; 1971b). This early work was quickly applied to fuzzy relational databases, image analysis, and decision support systems. It was not until Mamdani and his group of researchers (Mamdani and Assilian, 1975; Mamdani and Pappis, 1977) pioneered the use of fuzzy mathematics for control of laboratory scale problems, that the use of fuzzy logic for process control became a practical reality.

The early work in fuzzy process control by Mamdani's group was focused on the control of a small steam engine in a laboratory environment. This work paved the way for large-scale process control methods. The first large-scale industrial application of fuzzy control was to a cement kiln (Holmblad and Ostergaard, 1982). The development of the cement kiln controller was facilitated by the availability of an operator's handbook containing rules for controlling the system. This work of Holmblad and Ostergaard laid the foundation for subsequent rule-based fuzzy process control systems. Particularly, it represents the starting point for the research described in this book. We were attracted to these fuzzy systems because they were simple and linguistic in nature; we could control nonlinear systems with simple linguistic rules. A fundamental flaw with this control system remained. It was not adaptive; it could not change when the process changed.

Early on in the fuzzy process control effort, researchers discussed primitive adaptive fuzzy control systems. Procyk and Mamdani (1979) extended their initial fuzzy control system to form a self-organizing process controller that used a linguistic description of the control strategy as a basis for adaptation. It implemented a derivative-based search to discover effective linguistic rules. The research effort presented in this book extends this initial effort in adaptive control by employing new and developing tools from the field of computational intelligence including genetic algorithms and neural networks. We felt that we could use adaptive linguistic systems to learn very precise control rules and that human operators might accept such linguistic controllers that use the kind of linguistic rules they understand better than they have accepted mathematical "black boxes." However, we are getting a little ahead of ourselves.

1.3 WHERE WE GOT INVOLVED

Our initial thesis was that traditional control techniques were unnecessarily complex and that expert systems could be used to implement the strategies employed by effective operators. However, as we discovered, traditional expert systems were not the answer by themselves.

Our initial expert system was a process control advisor for a phosphate flotation plant (Davis, Jordan, and Stanley, 1990). (Chapter 16 addresses column flotation. Flotation is a process through which economically valuable minerals are separated in a water suspension from sand sized valueless material.) This prototype expert system was developed to systematically offer advice on the control of a commercial Florida phosphate flotation plant operation. The PC-based system was designed to receive input from remote sensors for circuit parameters such as feed rate, reagent addition, and pH and also accept input of operator supplied information. This information described the quality of the concentrate and tailings in each of the coarse, fine, and amine flotation circuits. The computer evaluated the information using rules obtained from industry representatives and Bureau of Mines flotation experts. The expert system then recommended actions to the operator based on problems identified in the flotation circuit. This expert system was designed to provide more uniformity among the flotation operators and to serve as a trainer for new operators. In addition, it represented an initial step toward fully automated systems for phosphate flotation.

As effective as the phosphate flotation advisor was, it did have some inadequacies. Most notably, the traditional production rules it used were not adequate to accomplish accurate control. To illustrate this shortcoming, let's consider an example situation in which non-fuzzy, if-then, production rules are used.

Consider the task of capturing the characteristics of a simple system represented by the function shown in Figure 1.1. This figure can be thought of as presenting a relationship between input and output parameters, and captures the essence of the task associated with both process control and modeling. The goal in control is to provide a desired value of the output variables based on the current values of the input variables.

Interestingly enough, this figure also represents the modeling problem. In a computer modeling problem, the goal is to use independent variables to determine dependent variables. When there is but one independent and one dependent variable, Figure 1.1 is sufficient. However, as in the control problem, when there are multiple input and output parameters, the curve in Figure 1.1 must be expanded into a higher dimensional surface. Both control and modeling problems can be thought of as surface fitting problems.

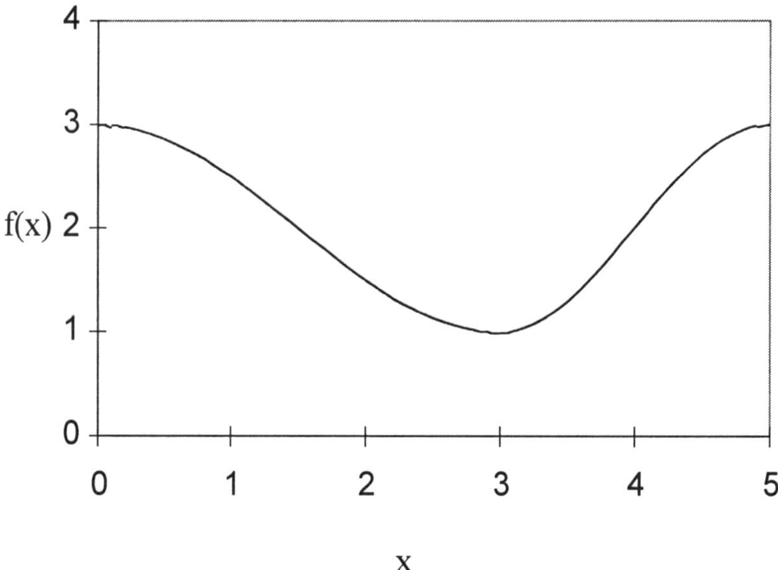

Figure 1.1 - A general functional relationship can be used to represent both process control and modeling applications.

Consider the task of representing the control/modeling problem represented by the curve in Figure 1.1 with production rules. And, realize that although this curve exists, it is generally identified with data points; we do not know the form of the equation that describes f(x). One production rule representation of the curve might consist of the following rules:

1. If $\{ 0 < x \leq 1 \}$ then $\{ f(x) = 3.0 \}$
2. If $\{ 1 < x \leq 2 \}$ then $\{ f(x) = 2.0 \}$
3. If $\{ 2 < x \leq 3 \}$ then $\{ f(x) = 1.0 \}$
4. If $\{ 3 < x \leq 4 \}$ then $\{ f(x) = 2.0 \}$
5. If $\{ 4 < x \leq 5 \}$ then $\{ f(x) = 3.0 \}$

The result of applying these rules is shown in Figure 1.2. Notice that the original real-valued function f(x) is represented by a step function f'(x). In some problem domains, this is an adequate representation; in most problem domains such an approximation of f(x) fails miserably. When a more accurate representation is required, the natural inclination is to simply add more rules:

1. If $\{ 0.0 < x \leq 0.5 \}$ then $\{ f(x) = 3.0 \}$
2. If $\{ 0.5 < x \leq 1.0 \}$ then $\{ f(x) = 2.5 \}$
3. If $\{ 1.0 < x \leq 1.5 \}$ then $\{ f(x) = 2.0 \}$

4. If { $1.5 < x \leq 2.0$ } then { $f(x) = 1.5$ }
5. If { $2.0 < x \leq 2.5$ } then { $f(x) = 1.0$ }
6. If { $2.5 < x \leq 3.0$ } then { $f(x) = 1.0$ }
7. If { $3.0 < x \leq 3.5$ } then { $f(x) = 1.5$ }
8. If { $3.5 < x \leq 4.0$ } then { $f(x) = 2.0$ }
9. If { $4.0 < x \leq 4.5$ } then { $f(x) = 2.5$ }
10. If { $4.5 < x \leq 5.0$ } then { $f(x) = 3.0$ }

The above rules result in the function f' (x) shown in Figure 1.3. Again, this finer representation of f(x) would prove adequate in some situations, yet inadequate in others. The approach of adding additional rules is not acceptable in most situations for at least three reasons: (1) there is a limit to the number of rules that can be used in order to meet real-time performance standards, (2) the function is still being represented discretely as a step function, and (3) people lose the feeling that the system is simple.

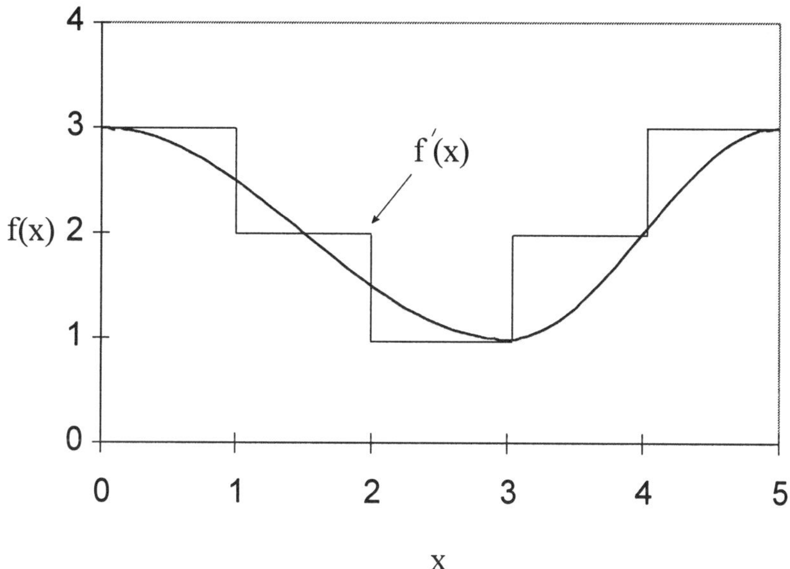

Figure 1.2 - f(x) is approximated with a step function when modeled using traditional if-then rules.

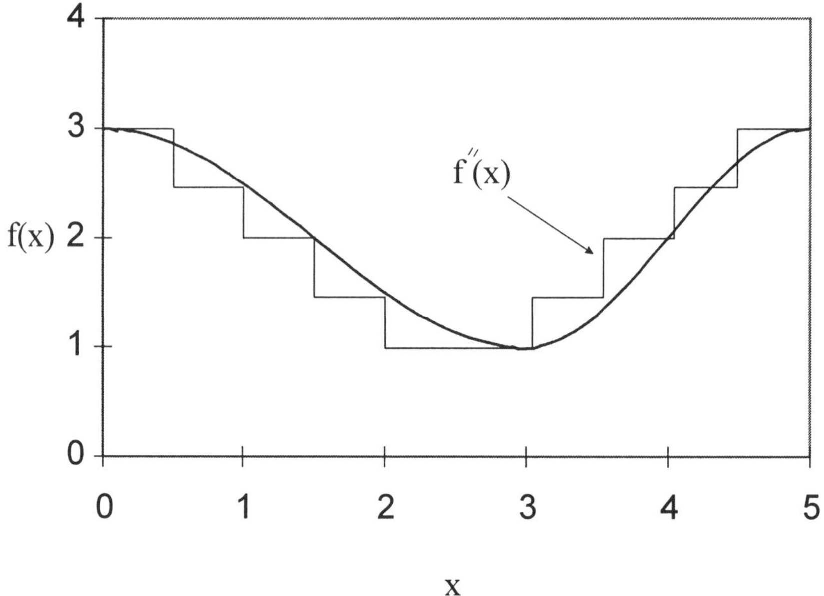

Figure 1.3 - Increasing the number of rules used to represent f(x) only makes the representation finer; it is still discrete.

The above discussion provides our major justification for shifting our focus from traditional production rules to fuzzy rules. The ability of fuzzy production rules to represent continuous functions in contrast to the discrete representation provided by the traditional production rules will be demonstrated in the next section. But first, consider a second rationale for not using traditional production rules.

As most any knowledge engineer will tell you, it is difficult to pin down the knowledge humans use in the form of rules. For the most part, human experts rarely have to think about the rules they use; the best ones generally work from their "feel" for the system. If a knowledge engineer is fortunate enough to get a human to write their rules, they are almost always in linguistic form and never in a numerical form that can be easily programmed using a computer. For instance, the five production rules producing f'(x) in Figure 1.2 might come from a human expert in the following form:

1. If { x is VERY SMALL } then { f(x) is BIG }
2. If { x is SMALL } then { f(x) is MEDIUM }
3. If { x is MEDIUM } then { f(x) is SMALL }
4. If { x is LARGE } then { f(x) is MEDIUM }
5. If { x is VERY LARGE } then { f(x) is BIG }

And these are the first fuzzy rules set forth in this book!

1.4 FUZZY PRODUCTION RULES FOR MODELING

In this section we provide the fuzzy production rules and the associated membership functions necessary to accurately reproduce (model) the mathematical function f(x) shown in Figure 1.1. This task will be accomplished using methods very similar to those set forth by Mamdani (1981). These methods are presented in a manner that should appeal to practicing engineers.

Let's begin with the five fuzzy production rules that appeared at the end of the previous section. The reader should understand the terms appearing in the rules above (VERY SMALL, SMALL, MEDIUM, etc.). However, a computer has absolutely no understanding of their meaning (computers work well with numbers, but poorly with linguistic concepts). Thus, to apply these linguistic rules, some meaning must be given to the fuzzy terms used to qualify the input variable x and the output variable f(x). For instance, under what circumstances is the first rule implemented? Well, it is implemented to some degree when x is VERY SMALL. Thus, we need some way for determining "to what extent" x is VERY SMALL. In fuzzy systems, this determination is made with fuzzy membership functions that provide linguistic terms with numerical (crisp) meaning.

Figure 1.4 shows fuzzy membership functions that can be used to describe the variable x. There are Gaussian representations for each of the five fuzzy terms used to describe x: VERY SMALL, SMALL, MEDIUM, LARGE, and VERY LARGE. Notice that for any value of x, the degree of membership, μ, for more than one term can be non-zero. The degree of membership can be thought of as the extent to which x is accurately represented by the term in question. If, for example, x is such that $\mu_{\text{VERY SMALL}} = 0.00$, then the interpretation is that the value of x is not described by the term VERY SMALL; it is not very small to any degree. If, on the other hand, x is such that $\mu_{\text{VERY SMALL}} = 1.00$, then the interpretation is that the value of x is exactly described by the term VERY SMALL. As an example, consider the situation in which x = 0.75. In this instance, the membership function values for the various fuzzy terms are: $\mu_{\text{VERY SMALL}} = 0.24$, $\mu_{\text{SMALL}} = 0.62$, $\mu_{\text{MEDIUM}} = \mu_{\text{LARGE}} = \mu_{\text{VERY LARGE}} = 0.0$. Thus, the variable value x = 0.75 has been *fuzzified* with the determination of the membership function values.

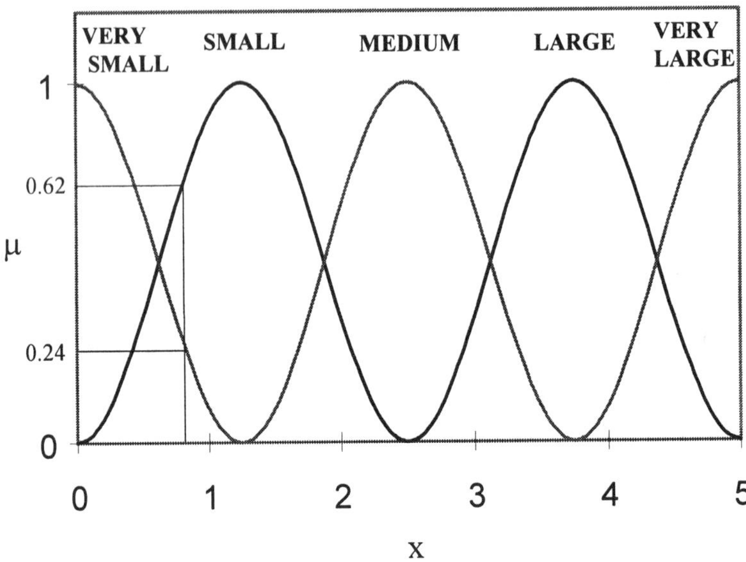

Figure 1.4 - Membership functions for the input variable x provide "meaning" to the fuzzy terms appearing in the linguistic rules.

Notice in Figure 1.4 that as x changes, so too do the values of the membership functions. Consider, for instance, what happens to the membership function values as x increases from 0.0 to 5.0. When x has a value of 0.0, then $\mu_{\text{VERY SMALL}} = 1.00$ and all other membership functions are 0.00. As x is gradually increased, the value of $\mu_{\text{VERY SMALL}}$ decreases and the value of μ_{SMALL} begins to increase (all other membership function values remain 0.00 at this point). We can think of this situation in the following way: as x is increasing, our *confidence* that x is VERY SMALL (or the extent to which we feel that value of x is accurately described by the term VERY SMALL) is waning. On the other hand, we feel that x is now to some extent described by the term SMALL. As x continues to grow, its membership in the set VERY SMALL goes down; its membership in the set SMALL goes up. With this representation, we do not have to decide that one single point at which x goes from being VERY SMALL to being SMALL. There is a continuous representation or mapping from discrete values to fuzzy sets, and yes, the sets overlap. We are no longer limited by conventional set theory in which "ball belongs to the set of red balls or the set of blue balls." Thank goodness, because in our world, there are definitely balls that are neither purely red nor purely blue, rather they are usually some shade of purple.

Now that x has been fuzzified, simply go to the rule set and implement the rule that is applicable. But wait one minute, more than one rule is applicable for a particular value of x. In fact, all of the rules that have nonzero membership function values are applicable to some extent, and this marks a major difference between traditional expert systems and fuzzy systems: in fuzzy systems more than one rule can be applied and there is no conflict between rules. In our particular example, rules 3-5 on page 7 really do not apply since x is not MEDIUM, nor is it LARGE, nor is it VERY LARGE (the membership function values for these sets is 0.00). Thus, these rules are thought of as being applicable to the degree zero. Rules 1 and 2, however, are characterized by fuzzy sets that have non-zero degrees of membership. The approach used to apply both rules in fuzzy systems is to take a weighted average of the value of f(x) prescribed by all applicable rules. Thus, the value of f(x) prescribed by rule 1 will be implemented to the degree 0.24, while the value of f(x) prescribed by rule 2 will be implemented to the degree 0.62.

The attempt to carry out the instructions of the previous paragraph, namely to implement the value of f(x) prescribed by the two applicable rules, i.e., to take a weighted average of the two rules, can bring to light another slight problem: How does one find the value of f(x) for BIG and MEDIUM, the values prescribed by the two applicable rules? To get values for the weighted average, membership functions are needed for the output variable f(x). Such membership functions are shown in Figure 1.5. Here, three membership function values are provided, although any number of membership functions could have been used.

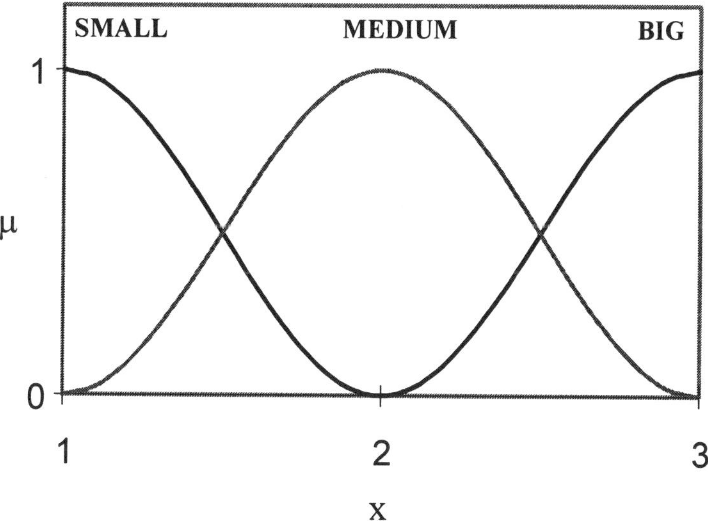

Figure 1.5 - Membership functions for the output variable f(x) are used to take weighted averages of applicable linguistic rules.

With the above membership functions defined, the process of taking the weighted average of the values of f(x) prescribed by the applicable rules can be thought of in graphical terms. What is done is to re-draw the membership functions associated with BIG and MEDIUM. However, the membership function forms are reduced or diminished according to the degree of membership associated with the particular rule. This process is depicted in Figure 1.6. BIG is reduced by cutting the function off at a height equal to the minimum value of μ associated with the condition side of the rule prescribing this action. In this particular instance, BIG is cut off at a height of 0.24 instead of the height of being drawn to a height of 1.00 as shown in Figure 1.5. MEDIUM is scaled with a height of 0.62. Now, the final crisp value of f(x) obtained for the given value of x = 0.75 is obtained by calculating the center of area (COA) of the two scaled-down membership functions as shown in Figure 1.6. The center of area is computed using the formula:

$$COA = \frac{\sum_{i=1}^{n} A_i \bar{x}_i}{\sum_{i=1}^{n} A_i}$$

where n is the number of rules, A_i is the area of the ith membership function describing f(x) for the ith rule, and x_i is the center of area for the membership function describing f(x) for the ith rule. The center of area for this instance provides the value of f(x) = 2.28; this is the final crisp value prescribed by the input x = 0.75. This *center of area* method is commonly used in fuzzy systems because of its simplicity. However, there are more involved defuzzification schemes and some will be used in later chapters of this book. But, in general, each method of defuzzification provides the controller with a mechanism for returning crisp (numerical) output values (f(x)).

This fuzzy system for computing a value of f(x) is relatively straightforward. It allows for the use of linguistic rules, it is easy to program on a computer, and it allows for an accurate representation of the function f(x). Figure 1.7 shows the function g(x) as defined by the fuzzy system described above. Notice that the function is not described by a series of steps as was the case with traditional production rules.

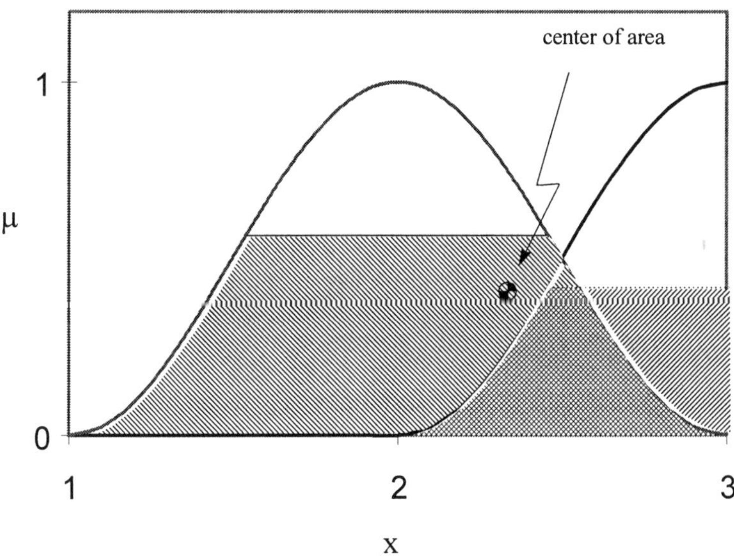

Figure 1.6 - Deffuzzification involves taking a weighted average of the truncated membership function values prescribed by applicable rules to compute a crisp value of f(x).

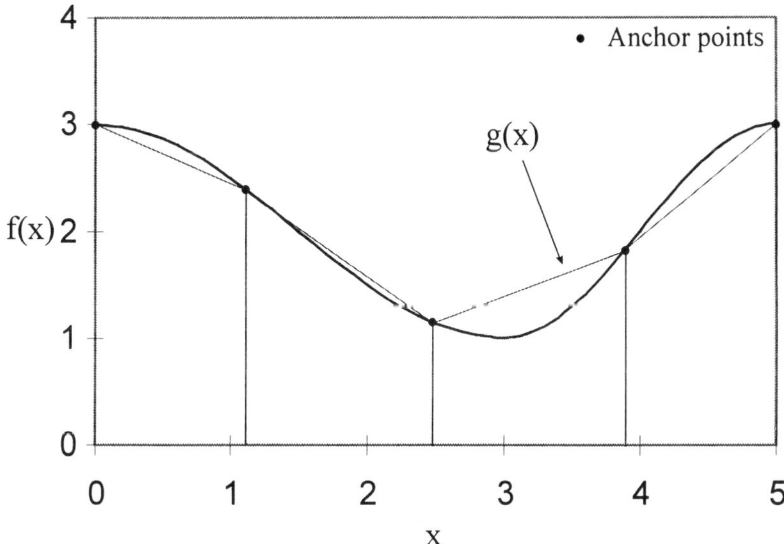

Figure 1.7 - g(x) is an approximation of f(x) using five fuzzy production rules.

1.5 REVIEW - PREVIEW

In this chapter we have attempted to do a few things. First, we gave a general synopsis of the reasons we chose fuzzy mathematics to serve as the core of our adaptive process control systems. For the most part, we felt the complex mathematics associated with traditional control schemes was unnecessary. Next, we provided some background as to how we came to view fuzzy systems as extensions of traditional expert systems. Finally, we provided an example of an instance in which a traditional production rule-based system was inadequate to solve a problem that could easily be solved using fuzzy production rules. The example provided was intentionally simple, but necessary because all of the complex applications that appear in this book are based on the fundamentals set forth in this chapter.

The development of the simple system that was provided quite possibly raised a number of questions for the reader. For instance, why were the membership functions Gaussian? How were the membership functions chosen? Are there other ways to take weighted averages? How does one handle more than one input? More than one output? How do I build one of these for a real problem? Where is the "adaptive" part? Unfortunately, it takes a good portion of this book to answer all of these questions (we don't get to the adaptive part until the second section, and we continue to build fuzzy systems all of the way through). However, we can begin to answer some of the questions, and we begin to do so in the next chapter. Chapter 2 presents a problem environment and an associated control problem. This problem environment is used to provide a step-by-step procedure for building fuzzy control systems; one that is used throughout the remainder of the book.

REFERENCES

Davis, B. E., Jordan, C. E. and Stanley, D. A. (1990). Expert advisor for phosphate flotation. *Control '90 - Society for Mining, Metallurgy, and Exploration*, 77-85.

Holmblad, L. P., and Ostergaard, J. J. (1982). Control of a cement kiln by fuzzy logic. In *Fuzzy Information and Decision Processes* (Eds. M. M. Gupta and E. Sanchez). Amsterdam: North-Holland Publishing Company, 389-399.

Larsen, P., M. (1980). Industrial applications of fuzzy logic control. *International Journal of Man-Machine Studies*, **12**, 255-283.

Mamdani, E. H. (1981). A fuzzy rule-based method of controlling dynamic processes. *Proceedings of the 20th IEEE Conference on Decision and Control*, December, San Diego, CA.

Mamdani, E. H., and Assilian, S. (1975). An experiment in linguistic synthesis with a fuzzy logic controller. *International Journal of Man-Machine Studies*, **21**, 213-227.

Mamdani, E. H., and Pappis, C. P. (1977). A fuzzy logic controller for a traffic junction. *IEEE Transactions on Systems, Man, and Cybernetics*, **SMC-7**, 707-717.

Melama, H., Barker, D., Hirvonen, M., and Hartikainen, J. (1987). Experiences gained from using an expert system approach on process management and control applications. *Proceedings of the Artificial Intelligence in Minerals and Materials Technology Conference*, October, Tuscaloosa, AL, 66-94.

Procyk, T. J., and Mamdani, E. H. (1979). A linguistic self-organizing process controller. *Automatica*, **15** (1), 15-30.

Tanimoto, S. L. (1987). *The elements of artificial intelligence.* Rockville, MD: Computer Science Press.

Zadeh, L. A. (1965). Fuzzy sets. *Information and Control*, **8**, 338-353.

Zadeh, L. A. (1968). Probability measures of fuzzy events. *Journal of Mathematical Analysis and Applications*, **23**, 421-427.

Zadeh, L. A. (1971a). Quantitative fuzzy semantics. *Information Sciences*, **3**, 159-176.

Zadeh, L. A. (1971b). Towards a theory of fuzzy systems. In: *Aspects of Networks and Systems Theory* (Eds., R. E. Kalman and R. N. deClairis), New York: Holt, Rinehart & Winston, 469-490.

Chapter 2

The First Fuzzy Controller

2.1 INTRODUCTION

It is our belief that engineers can produce effective fuzzy systems with minimal exposure to the theory behind fuzzy logic. Thus, in this chapter, we go directly to the heart of the issue by introducing a step-by-step procedure for developing fuzzy control systems. Details are provided only where they are needed; there are other references that provide mathematical details for interested readers. Initially, a simple liquid level system is introduced and used as a vehicle for introducing fuzzy control methods. When the reader understands the material presented in this chapter, he or she should be able to develop a rudimentary fuzzy control system.

2.2 THE PHYSICAL SYSTEM

The problem discussed in this chapter is the control of liquid level in a tank as shown in Figure 2.1. Figure 2.1a shows a schematic of the physical system and Figure 2.1b shows a block diagram of the control loop. The system includes an inflow and an outflow, both of which are adjustable using valves. For convenience, a mathematical model of the physical system is adjusted, not an actual tank containing liquid. (The controller that is developed has been applied to a scaled-down version of the simulated tank. The actual tank has digital sensors and pumps that accept digital control.) A tank having a total height of 50 m and a total cross sectional area of 78.53 m^2 is simulated. Both the inflow and outflow ranged between 0.0 m^3/sec and 200.0 m^3/sec. A time step of $\Delta t = 0.1$ sec is used. The physical system is governed by the continuity equation:

$$h^{t+1} = h^t + \left(\frac{Q_i - Q_o}{A}\right)\Delta t$$

where h is the liquid level height, A is the cross sectional area of the tank, Δt is the incremental time step used in the computer simulation, Q_i and Q_o are volumetric flow rates into and out of the tank respectively, and the superscript t indicates a particular time step.

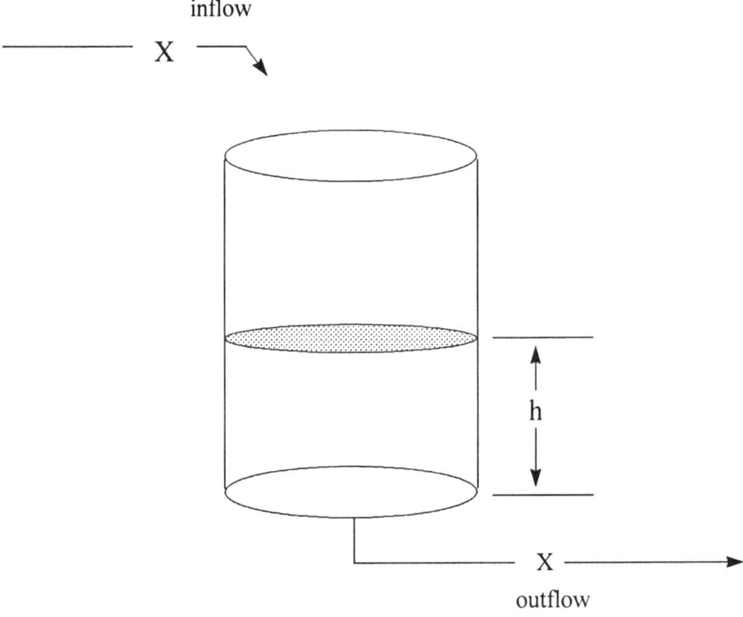

Figure 2.1a - The simple liquid level system includes both an inflow and an outflow. Note that we refer to the outflow as a negative inflow when defining fuzzy sets.

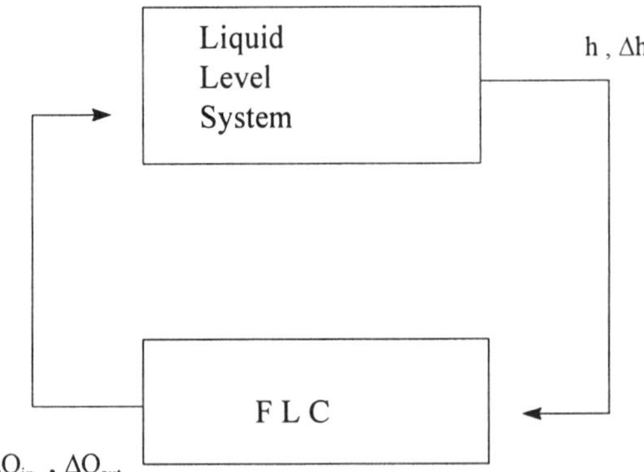

Figure 2.1b - The above block diagram demonstrates the control approach for the liquid level system. In this block diagram, h refers to the time-derivative of the height of the liquid level.

The model is used to develop a fuzzy logic controller (FLC) capable of driving the liquid level in the tank to a given setpoint. The liquid level and the rate at which this level is changing are initially set to arbitrary values. The control objective is to drive the system to a given setpoint in the shortest time possible by adjusting the inflow and the outflow. The setpoint is located at the centerline of the tank (h = 25 m). A constraint is placed on the motion of each valve so that no change in flow is allowed of over 20 percent of the current net flow rate. This constraint is enforced because in a physical system it would be necessary to limit transients in the pipes providing flow into and out of the tank.

It is not particularly difficult for a human to develop rules that when applied to the system drive the liquid level to a value that is relatively close to the setpoint. The rules for controlling the system are almost trivial: if the level is above the setpoint then remove liquid, if it is below the setpoint then add liquid. This liquid level system can be thought of as a simplified version of more complex systems that have been controlled using FLCs such as flotation cells (Shah, and Rajamani, 1988), casting plants (Bartolini, et al., 1985), and warm water plants (Kickert and Van Nauta Lemke, 1976). In fact, the FLC described in this chapter has served as the basis for controllers that manipulate real-world complex physical environments (Karr and Gentry, 1992).

2.3 DESIGN OF A FUZZY LOGIC CONTROLLER

There are numerous approaches to FLC development. Unfortunately, many of these approaches are complex and utilize cumbersome fuzzy mathematics. In this chapter, a straightforward approach to the development of an FLC is presented that simplifies things considerably. And this approach is used over and over again in this book to develop control systems for increasingly complex physical systems. A step-by-step procedure for fuzzy control of the liquid level system is provided. This procedure is written in a general form so that it can be easily adapted to the development of other FLCs. In discussing this procedure, special care is taken to relate the mechanics of the FLC to the methods a human might use to control the liquid level system. We think of our fuzzy systems as models of the actions taken by expert human operators (and later in the book we provide a mechanism to improve on the performance of the human operators, and even one to develop the requisite expertise from scratch).

The first step in developing the liquid level FLC is to determine the variables that will be important in choosing an effective control action. In the artificial intelligence community these variables are known as the *condition variables*. Two condition variables are readily identified as being important in the liquid level system. The control rules depend not on the current values of h and Δh (dh/dt), rather on the respective distances from the desired state, i.e., the setpoint. Therefore, the rules presented for the liquid level system use the following two variables, error (E) and change in error (ΔE) to be the condition variables:

$$E = h - h_s$$
$$\Delta E = \Delta h - \Delta h_s$$

where h_s is the desired, fixed setpoint for the level h, and Δh_s is the desired, fixed setpoint for the time rate of change of the level – both of which are to be driven to zero in our system.

Once the condition variables have been chosen, the *action variable(s)* must be identified. In this liquid level system, identifying the action variables is a straightforward task, for there are only two things the controller can adjust to alter the state of the liquid level system: either of the input or output flow rates (Q_i or Q_o) can be increased or decreased by adjusting a valve. To further simplify the problem, Q_i and Q_o can be considered together as a net rate of flow into the tank, Q_{net}, where:

$$Q_{net} = Q_i - Q_o .$$

In this problem, since Q_{net} was limited, the action variable was taken to be the percent change in Q_{net} (referenced using the variable ΔQ_{net}). In more complex systems, determining the important action variables is not always a straightforward task for there are often a large number of potentially important parameters that have complex relationships with the desired control result. Likewise, selecting the appropriate condition variables can be a daunting task because there are often a number of changes that, in the final analysis, really are of little or no consequence.

Once the important condition and action variables have been identified, the linguistic terms used to describe these variables must be defined (the *fuzzy sets*). In a classic computational intelligence approach to FLC development, fuzzy sets are written to describe the condition variables, E and ΔE. It must be understood that a greater number of linguistic terms can be expected to give more precise control. However, when more linguistic terms are used to describe the condition variables, more rules are needed by the FLC. (Recall, for instance, the expert system description of f(x) presented in Chapter 1 in which we improved the representation of f(x) by adding more rules.) Therefore, choosing the number of linguistic terms used to describe each variable is not an arbitrary decision; it is based on knowledge gained through working with the physical system.

For the liquid level system, based on the authors' experience with the liquid level problem, four fuzzy sets were chosen to characterize E: **Negative-Big (NB)**, **Negative-Small (NS)**, **Positive-Small (PS)**, and **Positive-Big (PB)**.

Five fuzzy sets were used to characterize ΔE: **Negative-Big (NB)**, **Negative-Small (NS)**, **Near-Zero (NZ)**, **Positive-Small (PS)**, and **Positive-Big (PB)**. Five fuzzy sets were used to characterize ΔQ_{net}: **Big-Positive (BP)**, **Small-Positive (SP)**, **No-Change (NC)**, **Small-Negative (SN)**, and **Big-Negative (BN)**. All fuzzy sets were chosen because they are similar to the descriptive terms a human operator might use to control the liquid level system.

This number of linguistic terms for each variable allowed for an FLC of reasonable size while providing adequate control of the liquid level system. With this choice of fuzzy sets 20 different production rules are required to describe all of the possible conditions that could exist in the liquid level system when the rules are of the form:

$$\text{IF [E is } \boldsymbol{A} \text{ and } \Delta\text{E is } \boldsymbol{B}\text{] THEN [}Q_{net} \text{ is } \boldsymbol{C}\text{]}$$

where ***A***, ***B***, and ***C*** are linguistic terms represented by fuzzy sets.

The method for giving precise meaning to the fuzzy, linguistic variables involves the use of *fuzzy membership functions*. The fuzzy membership functions used in the liquid level FLC are shown in Figures 2.2, 2.3, and 2.4. Fuzzy membership functions allow the crisp values of E and ΔE to be transformed into a fuzzy membership value.

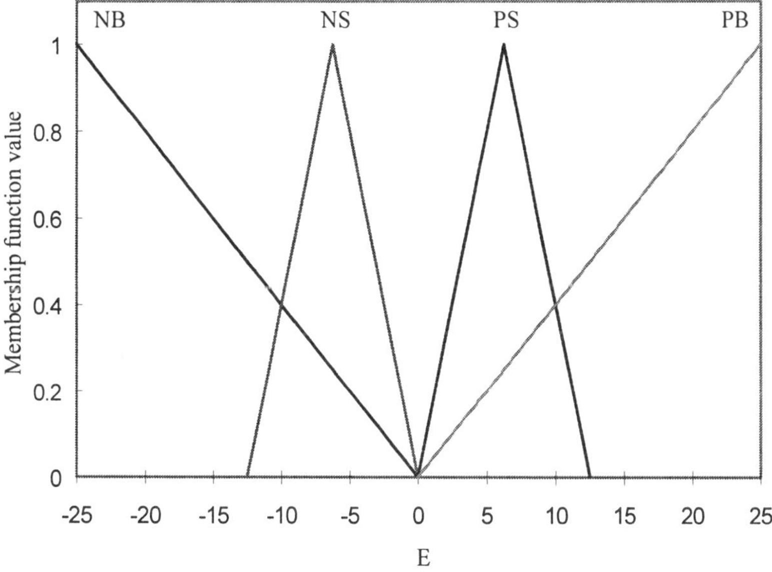

Figure 2.2. - Four fuzzy sets were used to describe error where NB is Negative-Big, NS is Negative-Small, PS is Positive-Small, and PB is Positive-Big.

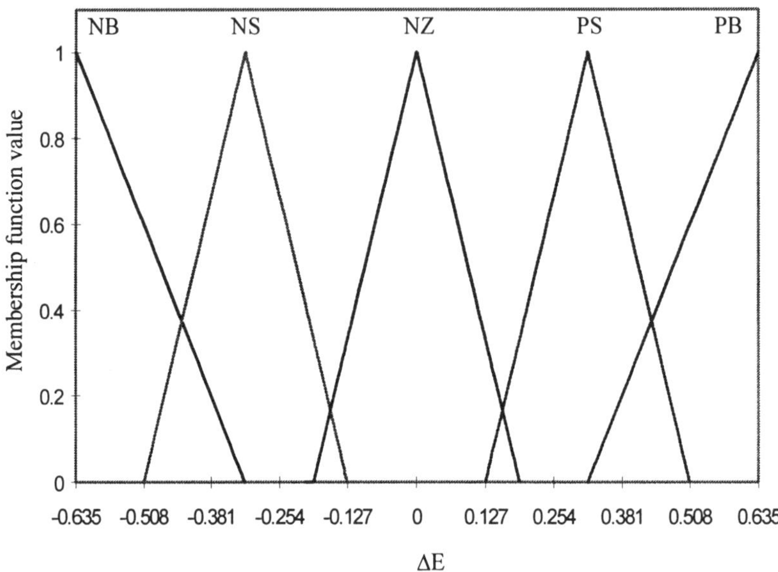

Figure 2.3. - Five fuzzy sets were used to describe time rate of change of error where NB is Negative-Big, NS is Negative-Small, NZ is Near-Zero, PS is Positive-Small, and PB is Positive-Big.

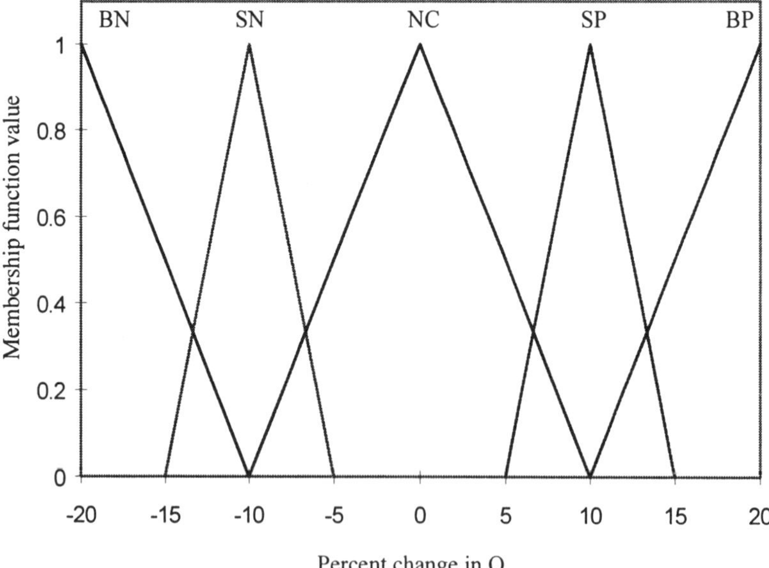

Figure 2.4. - Five fuzzy sets were used to describe percent change in flow rate where BN is Big-Negative, SN is Small-Negative, NC is No-Change, SP is Small-Positive, and BP is Big-Positive.

Actually, the fuzzy membership functions can be thought of as determining the degree to which a discrete value of a variable is described by a particular fuzzy, linguistic term.

For example, using the membership functions shown in Figure 2.2, a value of E=5 would produce the following membership values where μ represents a membership function value: μ_{NB}=0.0, μ_{NS}=0.0, μ_{PS}=0.8, and μ_{PB}=0.3. When a fuzzy membership function has a value of μ=1.0, the crisp value is accurately described by the fuzzy term. On the other hand, when μ=0.0, the crisp value is definitely not described by the fuzzy term. It is important to realize that for each crisp control value, each fuzzy set has a membership function value, even though some of the values are zero.

The crisp conditions (definite values of E and ΔE) existing in the liquid level system at any given time can be accounted for using production rules that include fuzzy linguistic terms. Now, a process for determining a crisp action to take on the liquid level system must be developed. In other words, the crisp conditions have been fuzzified (fuzzification). Now, they must be deffuzzified to yield a single crisp action. The set of fuzzy production rules provides a fuzzy action for any condition that could possibly exist in the problem environment. With the four fuzzy sets for E and the five fuzzy sets for ΔE, there are 4*5=20 possible conditions in the liquid level system, and a human expert provides a desirable action for each condition based on prior experience. An example of a fuzzy production rule used in the liquid level system follows:

IF [E is **PB** and ΔE is **PB**] THEN [ΔQ_{net} is **BN**].

This rule simply says that if the liquid level is well above the setpoint and rising rapidly, the net flow **into** the tank should be made negative big. The complete rule set for the liquid level FLC is depicted in Figure 2.5. This rule matrix is used by locating the descriptive term for E along the top of the matrix, locating the descriptive term for ΔE along the left side of the matrix, and then extracting the appropriate value for ΔQ_{net} for the given conditions.

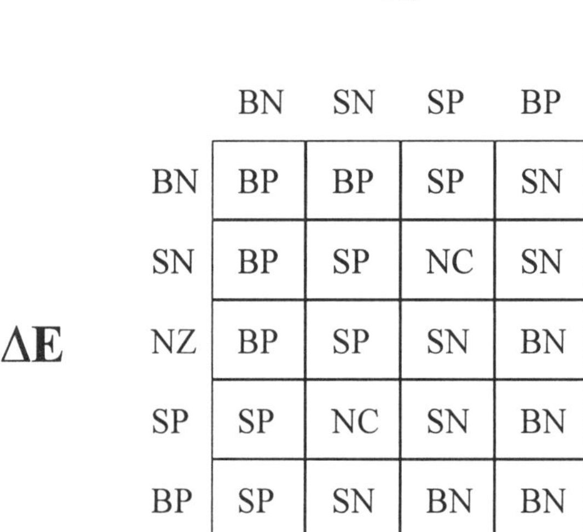

ΔE \ E	BN	SN	SP	BP
BN	BP	BP	SP	SN
SN	BP	SP	NC	SN
NZ	BP	SP	SN	BN
SP	SP	NC	SN	BN
BP	SP	SN	BN	BN

Figure 2.5. - The rule set included 20 rules that are summarized in the matrix above. The sample rule in the previous discussion – IF [E is **PB** and ΔE is **PB**] THEN [ΔQ_{net} is **BN**] – appears in the bottom right corner of the matrix.

Now, we are at that same point we arrived at in Chapter 1. Specifically, we have a means for converting a crisp set of measurements in the liquid level system to a set of fuzzy conditions, and a set of fuzzy production rules prescribing a fuzzy action associated with a particular set of fuzzy conditions have been developed. Unlike conventional expert systems where only one rule is applicable for any given set of conditions, all of the rules in an FLC take effect to some degree at every time step (albeit that we have zero confidence that some of the rules are appropriate as reflected by membership function values of μ=0). Therefore, there still remains the task of converting the 20 fuzzy actions prescribed by the fuzzy production rules into a single, crisp action to be taken on the liquid level system.

To review the approach we introduced in Chapter 1, Larkin (1985) found that a center of area (COA) method (sometimes called the centroid method) provides an efficient means for determining this crisp action. In the COA method, the fuzzy membership functions shown in Figure 2.4 are used in a weighted summing procedure to find one crisp action. These manipulated variable membership functions actually serve only to define the base points (those located on the membership function value axis) of the triangles because the heights of the membership functions are varied according to the membership function values associated with the individual rules as applied to a particular situation. Each production rule has its action (as described by a fuzzy membership function) scaled to the height equal to the minimum likelihood associated with the condition portion of the rule, i.e., the minimum μ value associated with the conditions under which the rule is to be applied.

For example, consider the sample rule provided above as applied when the conditions in the system are E=5 and ΔE=0.521. Since $\mu_{\mathbf{PB}}(E=5)=0.3$ and $\mu_{\mathbf{PB}}(\Delta E=0.521)=0.7$, the membership function for ΔQ_{net}=**BN** is "chopped off" at a height of 0.3. This is because the minimum membership function value for the variables on the condition portion of the rule is 0.3. When all 20 conditions have been compared with the rules, 20 trapezoids of varying height and width will have been plotted. Figure 2.6 shows a plot for the aforementioned condition existing in the environment: E=5 and ΔE=0.521. The two trapezoids occur as a consequence of the two rules that have their conditions met to some degree (the other 18 rules have a portion of their conditions met to the degree of 0.0 and are thus accounted for by trapezoids with a height of zero). The crisp action is determined simply by finding the center of area of the total area comprising the two trapezoids; in this case the crisp action is ΔQ_{net}=-12.0 percent. The COA is computed according to the following which is repeated from Chapter 1 for convenience:

$$COA = \frac{\sum_{i=1}^{n} A_i \bar{x}_i}{\sum_{i=1}^{n} A_i}$$

where A_i is the area of the triangle i, x_i is its COA, and n is the number of rules. The center of area of the composite shapes (they will be trapezoids and possibly triangles if any situation exists in which $\mu = 1$) is a straightforward calculation easily performed on a computer. This value defines the single crisp value of percent change in Q_{net} applied at the next time step. The rules that create triangles with large areas, those arising from the conditions that are most applicable, have the greatest effect on the action, as well they should.

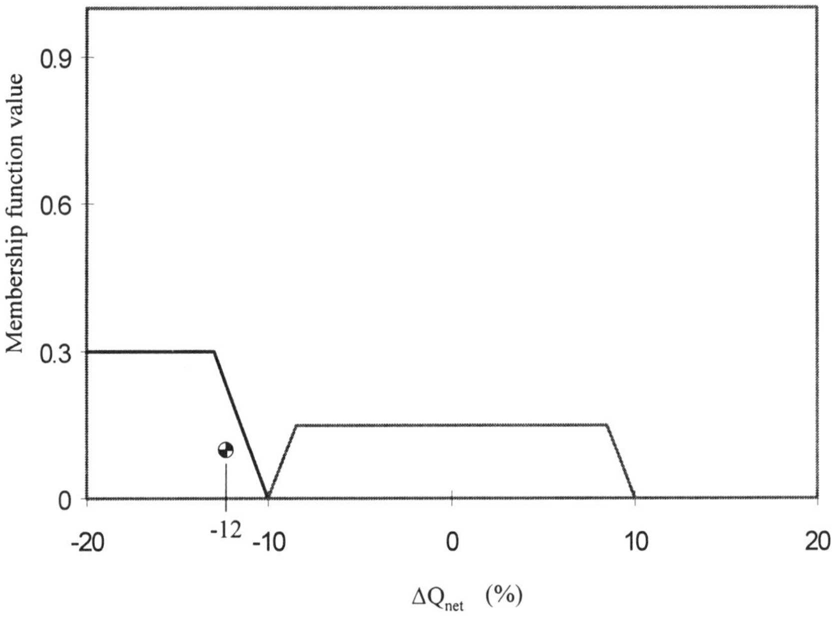

Figure 2.6. - The center of area method (sometimes called the centroid method) is a commonly used technique for defuzzification. In the above figure the COA of the two trapezoids occurs at ΔQ_{net} = -12.0.

The step-by-step procedure described above for developing an FLC is summarized below:

1) Determine the condition variables to be considered. These are the variables that are measured when a change to the system is considered.

2) Determine the action variables. These are the variables that are manipulated to elicit a change in the system state.

3) Describe the fuzzy sets for both the condition and action variables.

4) Establish fuzzy production rules that cover all of the possible conditions that exist in the problem environment.

5) Define the fuzzy membership functions.

6) Determine all condition membership values for each rule.

7) Apply the fuzzy production rules by taking a weighted average of the actions prescribed by all of the rules. The fuzzy production rules are

generally written in matrix form as shown in Figure 2.5 and the essence of the defuzzification process is depicted in Figure 2.6.

Certainly, this procedure is easily put in the form of a computer program.

To someone familiar with traditional control methods, this may seem like a strange approach to process control. To someone who has worked extensively with FLCs, this approach seems powerful, logical, and natural. At any rate, to help the reader become more comfortable with the effectiveness of this *fuzzy approach* to process control, consider the results from a computer program implementing the FLC described above that manipulates a mathematical model of the liquid level system. Figure 2.7 shows the liquid level height plotted as a function of time for one particular initial condition in the problem environment. The FLC uses only the set of 20 fuzzy production rules and the associated membership functions to govern its selection of actions. The FLC is able to drive the height to the setpoint of 25 m in approximately 70 sec at which time a small oscillation about the setpoint begins.

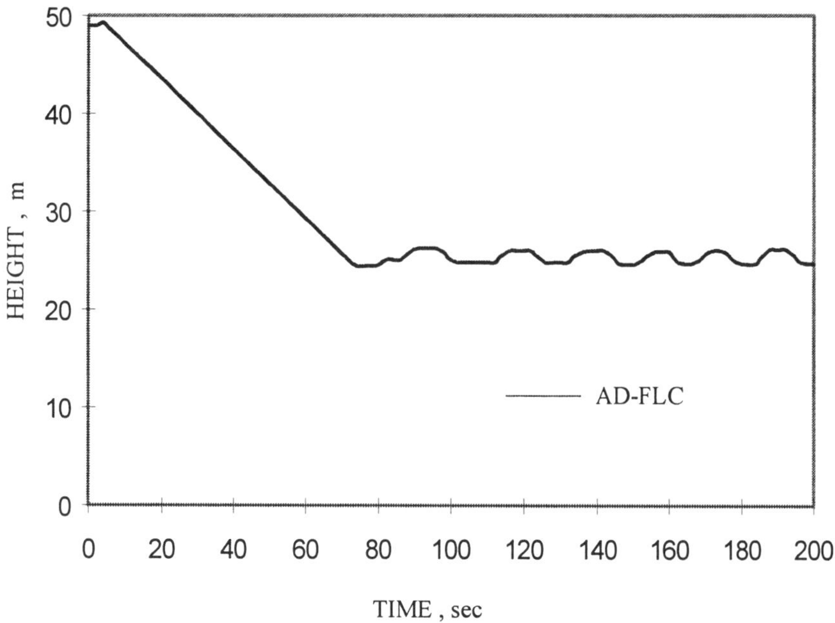

Figure 2.7. - The author-developed FLC successfully drives the liquid level to the setpoint.

2.4 REVIEW AND PREVIEW

This chapter has provided a step-by-step procedure for developing fuzzy control systems. This technique was demonstrated on a very simple liquid level system for which the reader could easily write some general, yet effective control rules. The idea of fuzzification via fuzzy sets was introduced (although hopefully it was done so subtly that you barely noticed it) along with the idea of defuzzification (simply taking a weighted average of the actions prescribed by the various rules). Results were provided to demonstrate that a control system could in fact be developed for the liquid level system using the guidelines presented.

When this material is presented to students, it seems to create as many questions as answers: (1) what if the rules are not readily apparent? (2) how do I get the membership functions? do their definitions affect the performance of the control system? (3) what if I don't have a computer model to develop my controller? and (5) what time is this class over? (there is always at least one in every group). Hopefully, you have some of the same questions and more (although the fourth question is not allowed from readers). Please, we will address these questions and plenty more we ran into during the course of the project. However, we feel there is now a more pressing need: to ensure that this simple step-by-step procedure for FLC development will work in more complex problem environments. Thus, in the next chapter we attempt to instill some confidence in this very straightforward approach by using it to develop a controller for a more difficult system, a cart-pole system that has become a standard in the computational intelligence-based controls community.

REFERENCES

Bartolini, G., Casalino, G., Davoli, F., Mastretta, M., Minciardi, R., and Morten, E. (1985). Development of performance adaptive fuzzy controllers with application to continuous casting plants, *Industrial Applications of Fuzzy Control* (M. Sugeno, ed.), North-Holland, Amsterdam, p. 73.

Karr, C. L., and Gentry, E. J. (1992). Real-time pH control using fuzzy logic and genetic algorithms. *Proceedings of Annual Meeting of the Society for Mining, Metallurgy, and Exploration*, Phoenix, AZ, February, 1992.

Kickert, W. J. M., and Van Nauta Lemke, J. R. (1976). Applications of a fuzzy controller to a warm water plant, *Automatica*, 12: 301-308.

Larkin, L. I. (1985). A fuzzy logic controller for aircraft flight control, *Industrial Applications of Fuzzy Control* (M. Sugeno, ed.), North-Holland, Amsterdam, p. 87.

Shah, I., and Rajamani, K. (1988). Fuzzy logic controller: Application to liquid level system. *Proceedings of IFAC Symposium on Automation in Mining, Metallurgy and Metals Processing*, Johannesburg, pp. 186-192.

Waterman, D. A. (1989). *A Guide to Expert Systems*, Addison-Wesley, Reading, MA.

Chapter 3

A Cart-Pole Fuzzy Controller

3.1 INTRODUCTION

In the previous chapter a step-by-step procedure for developing an FLC was introduced and applied to a very simple liquid level system. In this chapter, this same step-by-step procedure is used to develop a controller for a more difficult system, a cart-pole balancing system. The cart-pole system is more difficult for at least three reasons: (1) it has four condition variables, (2) it is a highly coupled system, and (3) the rules for manipulating the new system are not necessarily readily apparent. Nonetheless, the general approach produces a FLC that is capable of effectively controlling the cart-pole system.

3.2 CART-POLE BALANCING SYSTEM

In this simulated control problem a cart is free to translate along a one-dimensional track while a pole balanced on the cart is free to rotate only in the vertical plane of the cart and track. A force, F, which can take any one of a selected range of values in either direction along the track, is applied to the center of mass of the cart at discrete time intervals. No direct force is applied to the pole, thus any change in the motion of the pole is caused directly by the forces applied to the cart (and, of course, gravity). The objective of the control problem is to apply appropriate forces to the cart until it becomes motionless at the center of the track and to maintain the pole in a vertical position during the process. A controller for this cart-pole system must be able to achieve the control objective from a variety of initial conditions. This task of simultaneously centering a cart on a track while balancing a pole is often used as an example of the inherently unstable, multiple-output, dynamic systems present in many balancing situations, e.g., two-legged walking and stabilizing a rocket during vertical takeoff. A schematic of the cart-pole system is shown in Figure 3.1a. A block diagram of the control loop is shown in Figure 3.1b.

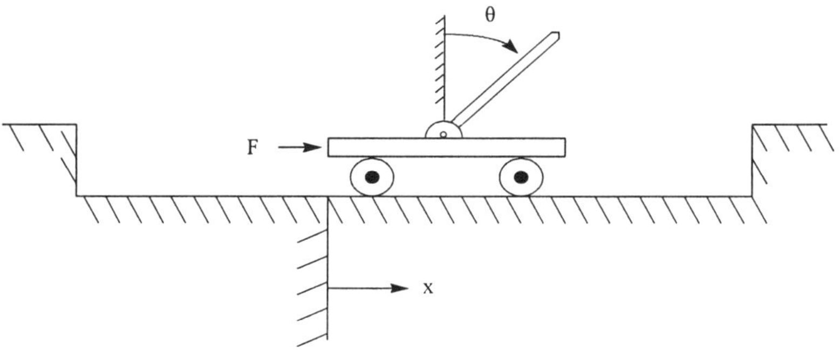

Figure 3.1a - Schematic of the cart-pole balancing system. Here, x is the position of the cart, θ is the angular position of the pole, and F is a force which can be applied directly to the cart.

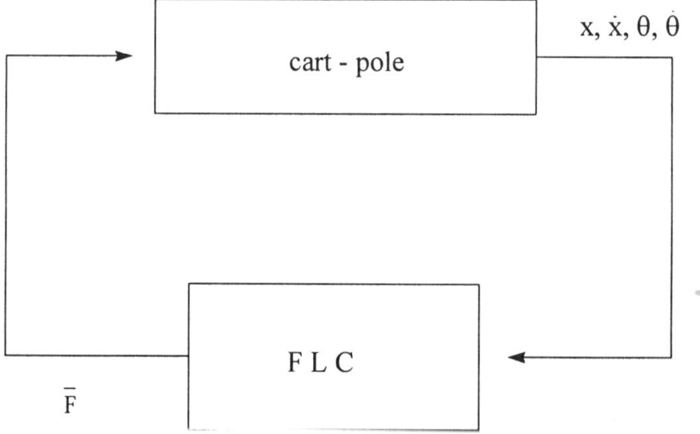

Figure 3.1b - Control loop for the cart-pole balancing system. Here, x is the position of the cart, $\dot{x}$ is the velocity of the cart, θ is the angle of the pole with respect to the vertical, $\dot{\theta}$ is the angular velocity of the pole, and F is a force applied to the cart.

The state of the cart-pole system at any time is described by four real-valued state or condition variables:

x = position of the cart;
$\dot{x}$ = velocity of the cart;
θ = angle of the pole with respect to the vertical;
and
$\dot{\theta}$ = angular velocity of the pole.

The system is modeled by the following nonlinear ordinary differential equations [2]:

$$\ddot{\theta} = \frac{g \sin\theta + \cos\theta \left[\dfrac{-F - m_p l \dot{\theta}^2 \sin\theta + \mu_c\, sign(\dot{x})}{(m_c + m_p)}\right] - \dfrac{\mu_p \dot{\theta}}{m_p l}}{l\left[\dfrac{4}{3} + \dfrac{m_p \cos^2\theta}{(m_c + m_p)}\right]} \tag{1}$$

$$\ddot{x} = \frac{F + m_p l [\dot{\theta}^2 \sin\theta - \theta \cos\theta] - \mu_c\, sign(\dot{x})}{(m_c + m_p)} \tag{2}$$

where the "double-dot" notation represents a second derivative with respect to time, and:

g = -9.81 m/s^2, acceleration due to gravity;
m_c = 1.0 kg, mass of cart (as will be discussed later, this value changes with time);
m_p = 0.1 kg, mass of pole;
l = 0.5 m, length of pole;
μ_c = 0.0005, coefficient of friction of cart on track;
μ_p = 0.000002, coefficient of friction of pole on cart;
and
-10.0 N $\leq$ F $\leq$ 10.0 N, force applied to cart's center of mass.

These equations indicate one of the major difficulties associated with controlling the cart-pole system: the system is coupled. Notice that the equation for the linear acceleration of the cart contains a term for the angular velocity of the pole, and the equation governing angular acceleration of the pole contains a term for the linear velocity of the cart. The values of cart and pole position depend on one another; the equations are coupled. Although this coupling complicates the solution of the equations, it allows both the cart to be centered and the pole to be balanced by changing only one variable, the applied force.

The solutions of the equations of motion were approximated using Euler's method of numerical integration (Press, et al, 1988), which yield the following difference equations:

$$\dot{\theta}^{t+1} = \dot{\theta}^{t} + \ddot{\theta}^{t} \Delta t \qquad (3)$$

$$\theta^{t+1} = \theta^{t} + \dot{\theta}^{t} \Delta t \qquad (4)$$

$$\dot{x}^{t+1} = \dot{x}^{t} + \ddot{x}^{t} \Delta t \qquad (5)$$

$$x^{t+1} = x^{t} + \dot{x}^{t} \Delta t \qquad (6)$$

where the superscripts indicate values at a particular time, Δt is the time step used in the numerical integration of the equations of motion, and the values $\ddot{x}^{t}$ and $\ddot{\theta}^{t}$ are evaluated using equations (1) and (2). A time step of 0.02 seconds was used to strike a balance between the accuracy of the solution and the computational time required to find a solution.

3.3 CART-POLE FLC

A FLC system presented in this chapter for control of a mathematical model of the cart-pole balancing system. The procedure necessary to accomplish this task was presented in the previous chapter, and it will be closely followed here. It is important to note that later in this book a more complicated version of the cart-pole system is addressed (one in which the mass of the cart will be a function of time); a version that is best handled using an adaptive controller.

The first step in developing the non-adaptive FLC considered here is to decide on the condition variables. These are the variables the must be taken into account when determining what action to take on the cart-pole system. For the cart-pole system, there are at least four variables that must be considered when determining a value of force with which to apply to the cart: (1) the position of the cart, x, (2) the velocity of the cart, $\dot{x}$, (3) the pole angle, θ, and (4) the angular velocity of the pole, $\dot{\theta}$. These are the condition variables: (1) the distance from the cart position setpoint, $E_x = s_x - x$, (2) the distance from the pole angle setpoint, $E_\theta = s_\theta - \theta$, (3) the distance from the cart velocity setpoint, $\Delta E_x = s_x - \dot{x}$, and (4) the distance from the pole angular velocity setpoint, $\Delta E_\theta = s_{\Delta\theta} - \dot{\theta}$. In these terms, the values s_x, s_θ, s_x, and $s_{\Delta\theta}$ are the respective setpoints for the cart position, pole angle, cart velocity, and pole angular velocity. As the problem has been defined, the values of these setpoints are all zero.

There is only one control action (and therefore only one control variable) that can be taken on this system and that is to select and apply a force to the cart. Thus, the controller is charged with the task of selecting a value of F by considering only the values of E_x, E_θ, ΔE_x, and ΔE_θ.

The third step in the design of an FLC is to select linguistic terms to represent each of the condition and action variables. Realize at the outset that there is not a definite method of doing this; the number and definition of the linguistic terms is always problem specific, and their definitions should be consistent with the normal meaning of the terms. For this application, four linguistic terms were used to describe the error associated with the position of the cart, E_x. Three linguistic terms are adequate to represent the variables, ΔE_x, E_θ, and ΔE_θ. The manipulated variable, F, required seven linguistic terms for adequate representation. The specific linguistic terms which describe the condition and action variables follow:

E_x Negative Big (**NB**), Negative Small (**NS**), Positive Small (**PS**) and Positive Big (**PB**);

ΔE_x Negative (**N**), Near Zero (**NZ**), and Positive (**P**);

E_θ Negative (**N**), Near Zero (**NZ**), and Positive (**P**);

ΔE_θ Negative (**N**), Near Zero (**NZ**), and Positive (**P**);

F Negative Big (**NB**), Negative Medium (**NM**), Negative Small (**NS**), Zero (**Z**), Positive Small (**PS**), Positive Medium (**PM**) and Positive Big (**PB**).

(These linguistic terms are later "defined" with the membership functions shown in Figure 3.3.)

The fourth step in the design of a FLC is to develop a rule set. The rule set in a FLC must include a rule for every possible combination of the variables of the chosen linguistic terms. Thus, 108 rules are required for the cart-pole balancing FLC. There are 4 * 3 * 3 * 3 = 108 possible combinations of the fuzzy linguistic terms describing the variables E_x, E_θ, ΔE_x, and ΔE_θ. Many of the actions needed for the 108 possible condition combinations are readily apparent.

For instance, when the position of the cart is **NEGATIVE BIG** (well to the left of center), the velocity of the cart is **NEGATIVE** (the cart is moving further still to the left of center), the position of the pole is **POSITIVE** (the pole is leaning back to the right), and the angular velocity of the pole is **POSITIVE** (the pole is moving further to the right), the best action is to apply a positive big force to the cart, i.e., to hit the cart hard on its left side. Such a force will tend to move the pole back to a vertical position while simultaneously moving the cart toward the center of the track. However, because the system is coupled (the pole is sensitive to actions taken on the cart), there are some conditions for which the appropriate action is not readily apparent. In fact, there are some conditions for which the selection of an appropriate action seems inappropriate.

For example, what is the appropriate action when the position of the cart is **PS**, the velocity of the cart is **P**, the position of the pole is **NZ**, and the angular velocity of the pole is **P**? The cart is to the right of the centerline and moving further away from the setpoint. Thus, if one considers only the cart, the appropriate action is to apply a small force in the negative direction. However, obviously the cart cannot be considered independently of the pole. The state of the pole requires a small force in the positive direction. The traditional way to resolve these conflicts and to select an appropriate action has been to experiment with different selections of the action variables. (As will be presented in several later chapters, an alternative to the trial-and-error approach is to allow a search algorithm like a genetic algorithm to select an effective rule set). However, at this juncture, the method of choice is the old fashioned trial-and-error approach directed by experience with controlling the system.

The complete rule set used for the cart-pole balancing system appears in Figure 3.2. To help ensure that the reader understands this figure, the bold action in the figure is the appropriate action for the condition: E_x is **PS**, ΔE_x is **NZ**, E_θ is **P**, and ΔE_θ is **N**. Note that the linguistic terms for E_θ are read as **N**, **NZ**, and **P** from left to right, and the linguistic terms for ΔE_θ are read as **N**, **NZ**, and **P** from top to bottom. Thus, the complete rule is:

IF {E_x IS **PS** AND ΔE_x IS **NZ** AND E_θ IS **P** AND ΔE_θ IS **N**} THEN {F IS **NM**}.

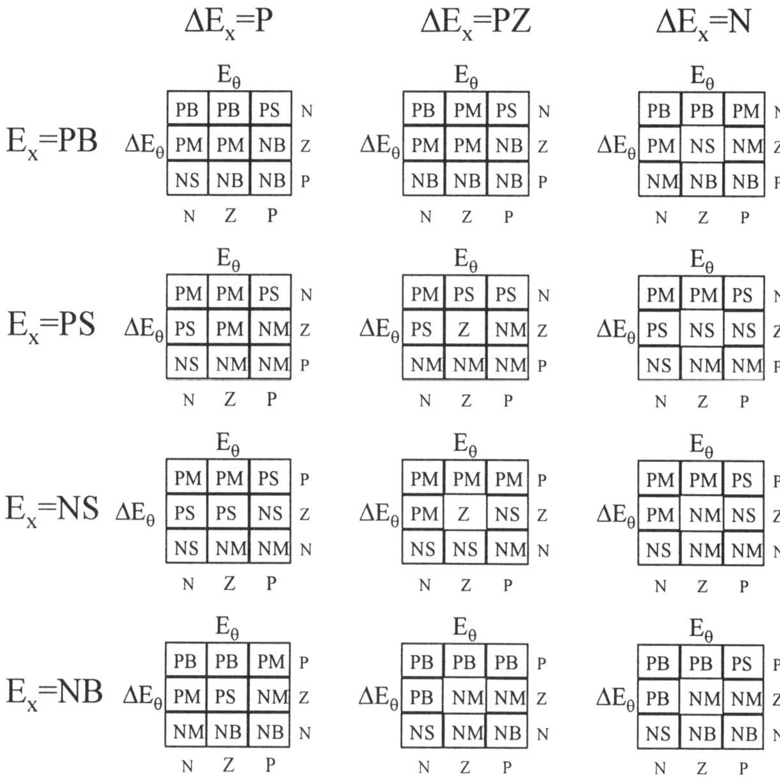

Figure 3.2 - The complete rule set for the cart-pole balancer includes 108 rules.

Now that the fuzzy rules are established, the exact criteria for determining which rules apply at any given time must be determined. In other words, the fuzzy terms must be "defined" through the membership functions for both the condition and action variables. As with the initial requirement of selecting the necessary linguistic terms, there are no definite guidelines for constructing the membership functions; the terms are defined to represent as closely as possible the ordinary linguistic meaning of the terms. Membership functions can have many forms. Triangles and trapezoids are two commonly used membership function forms (and the two forms that will be used in the applications presented in this chapter). A restriction generally applied to the membership functions is that they have a maximum value of 1 and a minimum value of 0. When a membership function value is 1, there is complete confidence in the premise that the crisp value of the condition variable is accurately described by the particular linguistic term. When a membership function value is 0, there is complete confidence in the premise that the crisp value of the condition variable is ***not*** described by the particular linguistic term. The membership functions developed by the authors for the cart-pole balancer appear in Figure 3.3.

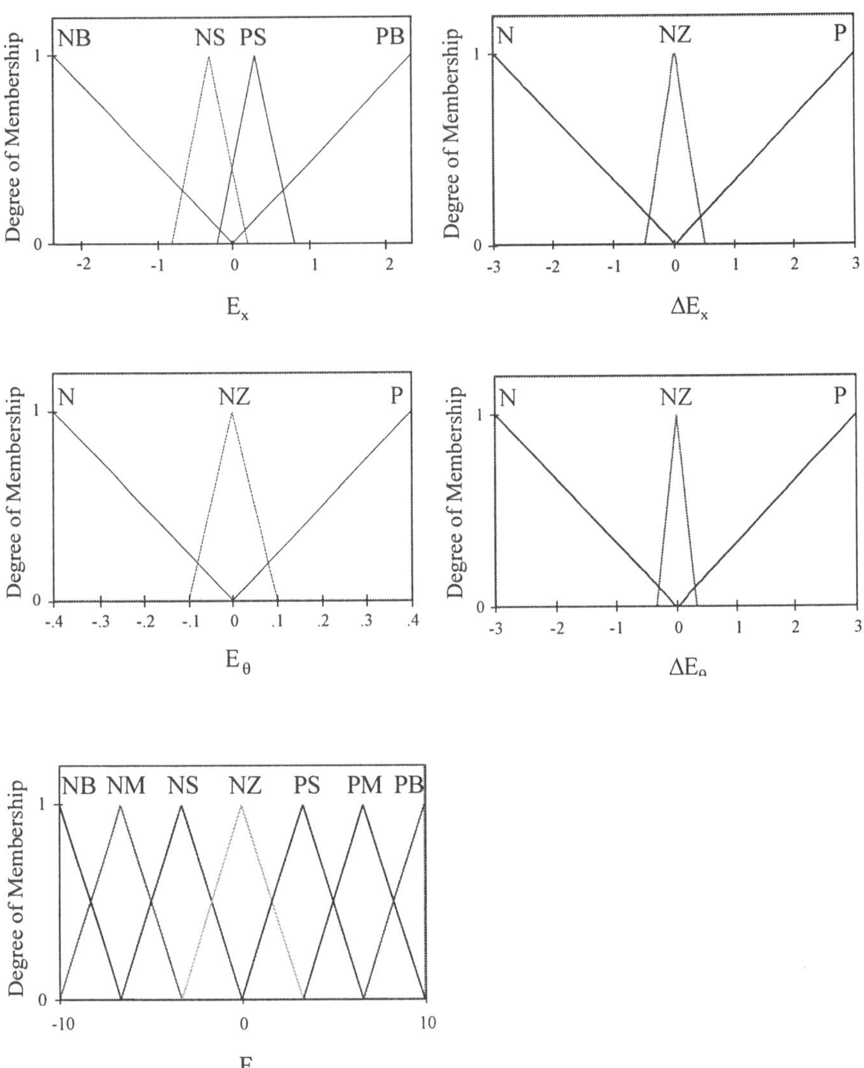

Figure 3.3 - Fuzzy membership functions used in the cart-pole balancer FLC.

Now that both the condition and action variables have been chosen and described with linguistic terms, and a rule set has been written that prescribes an appropriate action for every possible set of conditions, it is left to determine a single crisp value of the force to be applied to the cart at a particular time step. This is a concern because more than one of the 108 possible rules can be applicable for a given state of the cart-pole system.

A common technique for accomplishing this task is the Center of Area (COA) or centroid method (Larkin, 1985). In the COA method, the action prescribed by each rule can play a part in the final crisp value of F. The contribution of each rule to the final value of F is proportional to the minimum confidence (the minimum value of the membership function values on the left side of the rule) one has in that rule for the specific state of the physical system at the particular time. The net result is that all of the rules are considered in proportion to the confidence associated with the truth of the particular rule at the particular time. With the determination of a strategy for resolving "conflicts" in the actions prescribed by the individual rules, the FLC is complete.

In Chapter 2, the defuzzification involved "chopping off" triangles to obtain trapezoids. In the cart-pole balancer, an alternative approach is introduced in which the membership functions prescribed by a particular rule are not chopped off, rather they are scaled down. Thus, actions described by triangular membership functions remain triangular in the solution plots. To demonstrate this alternative approach to defuzzification, and to reinforce the way an FLC works, consider the following possible state of the cart-pole system: $x = -0.5$, $\dot{x} = -2.0$, $\theta = -0.5$, and $\dot{\theta} = 2.0$. For this state, the condition variables have the following values:

$E_x = -0.5$, $\Delta E_x = -2.0$, $E_\theta = -0.5$, and $\Delta E_\theta = 2.0$. These variable values are fuzzified via the fuzzy membership functions to elicit the following values:

E_x $\mu_{NB} = 0.25$, $\mu_{NS} = 0.35$, $\mu_{PS} = 0.0$, and $\mu_{PB} = 0.0$;

ΔE_x $\mu_N = 0.75$, $\mu_{NZ} = 0.0$, and $\mu_P = 0.0$;

E_θ $\mu_N = 0.15$, $\mu_{NZ} = 0.42$, and $\mu_P = 0.0$;

ΔE_θ $\mu_N = 0.0$, $\mu_{NZ} = 0.0$, and $\mu_P = 0.73$

The above membership function values are used to determine the extent to which the rules are applicable. It turns out that the following four rules are applicable:

IF {E_x IS **NB** AND ΔE_x IS **N** AND E_θ IS **N** AND ΔE_θ IS **P**} THEN {F IS **PB**}
IF {E_x IS **NB** AND ΔE_x IS **N** AND E_θ IS **Z** AND ΔE_θ IS **P**} THEN {F IS **PB**}
IF {E_x IS **NS** AND ΔE_x IS **N** AND E_θ IS **Z** AND ΔE_θ IS **P**} THEN {F IS **PM**}
IF {E_x IS **NS** AND ΔE_x IS **N** AND E_θ IS **N** AND ΔE_θ IS **P**} THEN {F IS **PM**}

The solution plot produced for this situation appears in Figure 3.4. Notice that the triangles have simply been scaled so that their height is equal to the minimum confidence associated with the rule that prescribes them; they maintain their triangular shape. According to this figure, the fuzzy cart-pole controller responds to a state in which $x = -0.5$, $\dot{x} = -2.0$, $\theta = -0.5$, and $\dot{\theta} = 2.0$ with a force of magnitude 7.2.

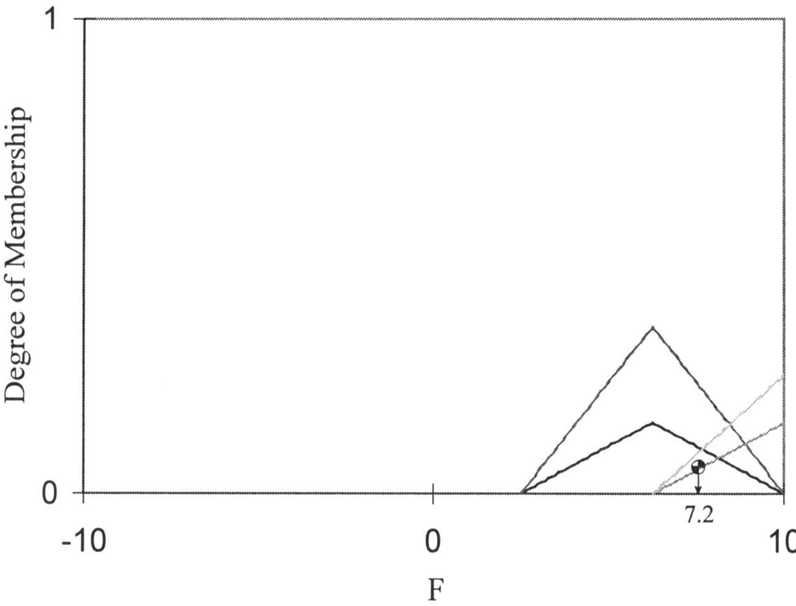

Figure 3.4 - Solution plot in which actions are obtained by scaling.

Figure 3.5 shows results for the cart-pole system described above. The rules and membership functions set forth by the authors form an FLC that effectively accomplishes the control task: it drives the cart to the center of the track in approximately 20 seconds while balancing the pole. This controller was developed over time through a trial-and-error process in which potential rules and membership functions were tested over and over again, until the ones shown were deemed adequate. However, as we will see later, a more effective controller can be developed with far less trial-and-error effort.

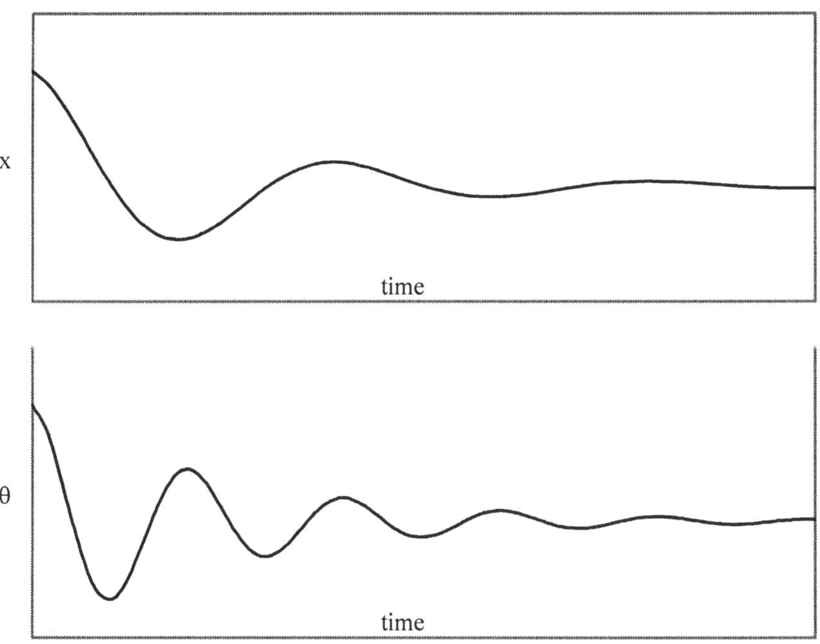

Figure 3.5 - The FLC is able to achieve the control objective of driving both the cart position and the pole angle to zero. Furthermore, the controller is able to keep these variables at their desired setpoints.

3.4 REVIEW/PREVIEW

In this chapter we have applied the step-by-step procedure set forth in Chapter 2 to a more difficult problem, a cart-pole balancing problem. The cart-pole system is more difficult for at least three reasons: (1) it includes four condition variables, (2) it is a coupled system, and (3) the appropriate rules are not readily apparent. Nonetheless, an effective FLC has been developed that is capable of controlling the system from a variety of initial condition positions.

This chapter also introduced an alternative defuzzification procedure. In Chapter 2 we presented a center of area approach that used triangular action membership functions that had been chopped off. In the current chapter we introduced an alternative means for scaling membership functions in which membership functions maintain their shape but are reduced in size. Initially, the authors experimented with this alternative approach because we were always looking for a better, more effective or more efficient way of doing things. Our experience (and it took us quite a while to convince ourselves of this) led us to the conclusion that it really did not matter which approach we used. We could get either approach to produce an effective controller. This effort is summarized in a report by Karr, Fleming, and Vann (1994).

At this point in our research, we felt that the successful development of a cart-pole fuzzy controller was a nice milestone. After all, we had developed a system that could use linguistic rules similar to those used by human operators to manipulate a system that is difficult to control. However, we also felt that we had a number of questions that needed exploring. Some of the questions were fairly obvious like, can this approach be used on a real problem? Are there better ways to select the rules; the membership functions? Some of the questions were not quite as obvious and required more subtle answers such as what happens when the system has additional parameters that change periodically such as the mass of the cart and the length of the pole? In the remainder of this book we will address these questions and others. However, it will take us a little time to get to them all. In the next chapter we will discuss an opportunity we were presented with at about this time in the project; we will discuss the development of a controller for a space satellite rendezvous system.

REFERENCES

Larkin, L. I. (1985). A fuzzy logic controller for aircraft flight control. *Industrial Applications of Fuzzy Control* (M. Sugeno, ed.), Amsterdam: North-Holland, p. 87.

Karr, C. L., Fleming, J. W., & Vann, P. A. (1994). *Aspects of genetic algorithm-designed fuzzy logic controllers* (Report of Investigations number 9515). Washington, DC: U.S. Department of the Interior, Bureau of Mines.

Press, W. H., Flannery, B. P., Teukolsky, S. A., and Vetterling, W. T. (1988). *Numerical recipes in C.* Cambridge, MA: Cambridge University Press, p. 567.

Chapter 4

A Satellite Rendezvous Fuzzy Controller

4.1 INTRODUCTION

Thus far, a step-by-step procedure for developing fuzzy control systems has been presented, and used to develop fuzzy controllers for a simple liquid level system and for a cart-pole balancing system. In the current chapter, this same procedure is used to develop a control system for satellite rendezvous. The rendezvous problem is important in the aerospace industry, and represents a step up in complexity from the cart-pole system for at least two reasons: (1) the equations of motion are again strongly coupled, but in three dimensions, and (2) there is more than one action that can be taken on the satellite.

The system consists of two orbiting spacecraft, one actively performing the rendezvous (the chaser), the other passively orbiting (the target). The rendezvous problem has been described in the literature (Natenbruk and Ragnittk, 1983) and is a challenging control problem. An FLC is developed that is capable of performing the rendezvous of the vehicles whose motions are simulated by a computer model.

This chapter introduces three new ideas. First, a new type of rule that uses a real-valued consequent is introduced and explained. This new rule type uses a real valued number to describe the action variables. Thus, in the real valued consequent approach the description of the action variables is not restricted to a finite number of linguistic terms as it was in the previous chapters. The real valued consequent approach therefore provides for a more robust control system. Second, the membership functions are manipulated using a multiplication operator rather than the minimum operator previously used. Third, a dual rule set is developed to account for coupling conditions found in the rendezvous system. Together, the real valued consequent approach, the multiplication operator, and the dual rule set provide the system developer with more power and flexibility. Increased power and flexibility are needed to control complex, real-time systems such as spacecraft rendezvous.

4.2 THE PHYSICAL SYSTEM

Autonomous spacecraft rendezvous operations are of importance to many space activities. This chapter is concerned with the "terminal" phase of the rendezvous maneuver that occurs just prior to docking. This terminal phase is the most difficult aspect of rendezvous to automate.

Control of the terminal phase of the rendezvous is accomplished using information from sensors onboard the chase vehicle that measure relative range and closure rate. Using a relative coordinate system in which all values are measured from the chaser vehicle position simplifies calculation. Such a coordinate system makes the target appear motionless to the chaser vehicle. In this study, the relative coordinate system (Figure 4.1) is fixed to the target vehicle with the y-axis directed radially from the earth, the x-axis tangent to the orbit in the direction opposite the target's motion, and the z-axis normal to the plane of the orbit. The z-axis is chosen to complete the customary right hand system.

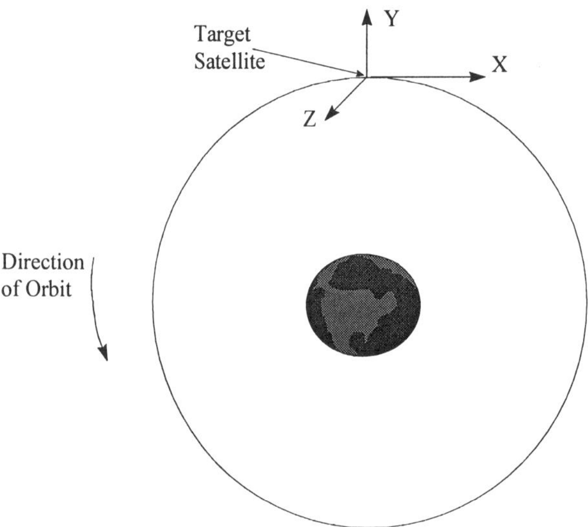

Relative Motion Coordinate System

Figure 4.1 - The relative coordinate system used in the spacecraft rendezvous problem makes the target appear motionless to the chaser vehicle.

The complete differential equations of relative motion for this rendezvous system are readily available (Kaplan, 1976). When the distance between the spacecraft is small compared with the distance from the target to the center of the earth, the three equations of relative motion reduce to the Clohessy and Wiltshire (sometimes called Hill) equations (Clohessy and Wiltshire, 1960):

$$\ddot{x} - 2a\dot{y} = \frac{F_x}{m}$$

$$\ddot{y} + 2a\dot{x} - 3\omega^2 y = \frac{F_y}{m}$$

$$\ddot{z} + \omega^2 z = \frac{F_z}{m}$$

where ω is the angular velocity of the relative coordinate system origin rotating about the earth; F_x, F_y, and F_z are thrust forces; m is the mass of the chaser vehicle; and x, y, and z are the relative coordinate positions.

In the present study the equations of motion provided above are numerically integrated using a predictor-corrector linear multistep method, specifically the explicit Adams-Bashforth 3-step method for the predictor and the implicit Adams-Moulton 3-step method for the corrector (Burden, Faires, and Reynolds, 1978). The computer model that results from the numerical integration of the above equations of motion is used to provide a fuzzy control system with position and velocity information that is provided by sensors in an actual rendezvous system.

4.3 A FUZZY PROCESS CONTROLLER

The first step in developing an FLC, using a real valued consequent and the multiplication operator for the spacecraft rendezvous system, is to determine which variables are important for effective control. Information about these variables must be available from sensors in the real-world system or from a computer model. An effective control decision can be made based on the position and velocity of the spacecraft relative to the target. Thus, there are six condition variables: (1) the x position coordinate, x, (2) the y position coordinate, y, (3) the z position coordinate, z, (4) the x-velocity, $\dot{x}$, (5) the y-velocity, $\dot{y}$, and (6) the z-velocity, $\dot{z}$. The state of the spacecraft rendezvous system at any time is completely described by these six variable values. These variables must appear in the fuzzy rules; they are not used to form error and change-in-error expressions as was done in the previous chapters. The main reason for working directly with the system parameters instead of converting them to error and change in error terms is that the human expert who was consulted on this problem was more comfortable working directly with the system parameters.

The second step in FLC development is to define the action variables, those variables that will be used to control the system. In the spacecraft rendezvous system there are three things that can be done to effect a change in the state of the system: (1) thrust applied in the x direction, F_x, (2) thrust applied in the y direction, F_y, and (3) thrust applied in the z direction, F_z. Recall that in previous chapters the systems being considered had only one action variable.

Once the relevant condition and action variables have been identified, the linguistic terms used to describe the six controlled variables must be defined. The same six linguistic terms were used to describe each of the six controlled variables: NEGATIVE-BIG (**NB**), NEGATIVE-SMALL (**NS**), NEGATIVE-CLOSE (**NC**), POSITIVE-CLOSE (**PC**), POSITIVE-SMALL (**PS**), and POSITIVE-BIG (**PB**). These fuzzy classes were chosen because they are similar to the descriptive terms a human operator might use when attempting to rendezvous the spacecraft. The fuzzy membership functions used in FLCs are often either triangular or trapezoidal. Triangular membership function forms are used in the spacecraft rendezvous FLC. The membership function forms for the position variables (x, y, and z) are shown in Figure 4.2.

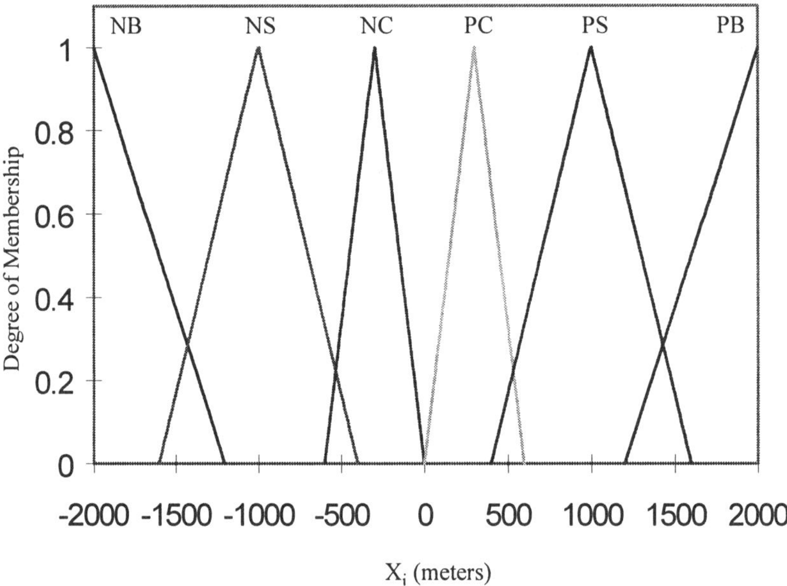

Figure 4.2 - The six triangular fuzzy membership functions for x (y and z) position(s) strike a suitable balance between timely computation and acceptable control.

The membership functions in Figure 4.2 are used to fuzzify crisp conditions (numerical values of the condition variables) existing in the spacecraft rendezvous system. Now, a process for determining a crisp action to take, an adjustment to each of the three thrusters, must be set forth. This process is based on a set of fuzzy production rules. Each fuzzy production rule provides a recommended fuzzy action for any possible condition that could possibly exist in the problem environment. The choice of fuzzy terms described above allows for the possibility of 36 different conditions that could exist in each of the x, y, and z coordinate directions when the rules are of the form (example for the x direction):

IF {x is α AND $\dot{x}$ x s β} THEN {F_x is γ}

where α and β are linguistic terms (or fuzzy classes) characterizing the respective variables, and γ is a precise numeric value for the thrust. (NOTE: The rule described here takes into account the x-direction only. The other directions are accounted for independently as will be described shortly.)

In the fuzzy rules used in previous chapters, and in most traditional FLCs the action variables, like the condition variables, are prescribed using one of a discrete number of fuzzy sets that describe the action to be taken on the system. For instance, if F_x is described with seven fuzzy terms (for example NEGATIVE BIG, NEGATIVE MEDIUM, NEGATIVE SMALL, NEAR ZERO, POSITIVE SMALL, POSITIVE MEDIUM, and POSITIVE BIG), then for each combination of condition variables, the developer has seven values from which to choose the value of F_x. This is an inviting format for rules in that the resulting rules are entirely linguistic, both the conditions and the actions. However, it does limit the flexibility associated with the development of the controller; there are only seven terms from which to choose the action variable.

In the rule form described above and used in the rendezvous controller, there is no limit within an appropriate range on the choice of action variable; F_x, F_y, an F_z can be represented by floating point numbers and not integer values. Because of the great degree of accuracy that is required, having a limited number of values from which to choose the actions would severely restrict the effectiveness of the controller. Thus, a real-valued consequent approach has been adopted. In this approach, the value of each consequent, i.e., the value of the thrusts, is a real number. Thus, since these real numbers can be of any precision desired, the developer actually has an infinite number of choices for each action. The action is no longer chosen from a limited number of values, rather from a limited range of values. Because of this increased flexibility, finer control can be achieved.

Actually, the real number that is used to prescribe the fuzzy action can be thought of as being represented by a triangular membership function. When this view of the real-valued consequent approach is adopted, the real number simply indicates the center of area of an isosceles triangle that has a base width of 2.0 (thereby yielding an area of 1.0). In this way, the center of area approach to defuzzification can still be used. The formula

$$COA = \frac{\sum_{i=1}^{n} A_i \bar{x}_i}{\sum_{i=1}^{n} A_i}$$

still applies, only now x_i is the real valued consequent and $A_i = \mu_i * (1.0)$. The only difference is that the developer is not limited by a finite number of fuzzy actions, and therefore can develop systems of far greater precision than with a more traditional fuzzy logic approach. When real numbers are used to characterize the actions, initial indications are that the linguistic nature of the fuzzy controllers is lost. However, it is not difficult to go back and "fuzzify" the real values describing the actions using fuzzy terms; its just that now there are potentially as many terms describing the action variable as there are rules.

In the example rule provided above, the coordinate directions are considered separately. However, the coordinate directions should by no means be considered separately — the system is highly coupled and applications of thrust in the x direction will affect motion in all three directions (look back at the equations of motion — x depends on y and y depends on x). However, if rules were written for each possible combination of the controlled actions, the result would be $6^6 = 66656$ rules. In general, the coupling effects need only be considered as the spacecraft approaches the target. Therefore, the coordinate directions are considered separately as long as the spacecraft is more than some predefined distance from the target (requiring 3*36=108 rules — 36 for each coordinate direction). Once the spacecraft is within this predefined distance, a set of "coupled rules" is implemented. These rules are of the form:

IF <x is α AND y is β AND $\dot{x}$ is γ AND $\dot{y}$ is Δ> THEN <F_x is δ AND F_y is ε>

where α, β, γ, and Δ are fuzzy terms defined by membership functions like those shown in Figure 4.2 and δ and ε are crisp numeric values. These rules are written for all of the possible combinations of the terms when the four relevant decision variables are in the NEGATIVE-CLOSE or POSITIVE-CLOSE classes. Thus, there are 16 coupling rules that, when combined with the 108 original rules, provide 124 total rules.

At this point a means has been developed for converting a crisp set of conditions existing in the spacecraft rendezvous system to a set of fuzzy conditions, and a set of fuzzy production rules prescribing a real-valued action associated with a particular set of fuzzy conditions. There still remains the task of converting the 124 real-valued actions prescribed by the fuzzy production rules into a single, crisp action to be taken on the rendezvous system. As in previous chapters, the COA method is used to combine all of the applicable rules. In the case of the real-valued consequent approach, the COA method simply yields the average of the action values (real-values) weighted by the product of the confidence levels. It is important to note here that the product of the confidence levels for the six controlled variables were used, not the minimum values. The use of the multiplication operator provided the same type of smooth mappings generated using the minimum operator. However, this is an alternative approach that is sometimes used and thus should be introduced to the reader.

4.4 DETAILS OF THE RENDEZVOUS CONTROLLER

This chapter has introduced two new elements to our fuzzy controllers: a real-valued consequent approach and a multiplier operator. Since these new ideas (especially the real-valued consequent approach) are potentially confusing, it is worthwhile demonstrating the details of an abbreviated example by hand.

To consider the details of a single iteration through the rendezvous controller, we first need to define the membership functions for the condition variables, and this has been done as shown in Figure 4.2. Next, we need a rule set. A complete rule set would consist of 124 rules. However, since the intent here is simply to demonstrate the procedure and explain the details of the operation, only the rules associated with F_x will be considered. These rules consist of the 36 rules that are independent of the y and z coordinate directions, and the 16 coupled rules. The abbreviated rule set is shown in Figure 4.3 in which the values of F_x have been normalized; similar rule sets exist for the other two coordinate directions.

x \ $\dot{x}$	NB	NS	NC	PC	PS	PB
NB	1.00	0.85	0.60	0.40	0.20	0.10
NS	0.85	0.60	0.40	0.20	0.10	0.00
NC	0.20	0.10	0.00	-0.10	-0.20	-0.40
PC	-0.20	-0.10	0.00	0.10	0.20	0.40
PS	-0.85	-0.60	-0.40	-0.20	-0.10	0.00
PB	-0.10	-0.20	-0.40	-0.60	-0.85	-1.00

Figure 4.3 - The above abbreviated rule set is applicable to the x direction only — there exist similar rule sets for the y and z directions.

Now, consider a specific state of the rendezvous system: x = -1525 and $\dot{x}$ = 0.0. The general procedure is to fuzzify the condition variables, apply the rules, and then defuzzify the actions. For fuzzification, we use the sensory output from the system to determine the appropriate membership function values:

x $\quad \mu_{NB} = 0.24$, $\mu_{NS} = 0.25$, $\mu_{NC} = 0.0$, and $\mu_{PC} = 0.0$, $\mu_{PS} = 0.0$, $\mu_{PB} = 0.0$;

$\dot{x}$ $\quad \mu_{NB} = 0.0$, $\mu_{NS} = 0.0$, $\mu_{NC} = 0.10$, and $\mu_{PC} = 0.10$, $\mu_{PS} = 0.0$, $\mu_{PB} = 0.0$;

Next, these membership function values are used to determine the extent to which each rule is applicable (both the independent and the coupling rules). Here, the minimum operator is dismissed in favor of a multiplication operator. The multiplication operator simply takes the product of the membership function values on the condition side of each rule. Figure 4.4 shows the abbreviated rule set with the corresponding weighting factors that are the product of the associated membership function values. Notice that for the above state, there are only four rules that are applicable to a non-zero degree.

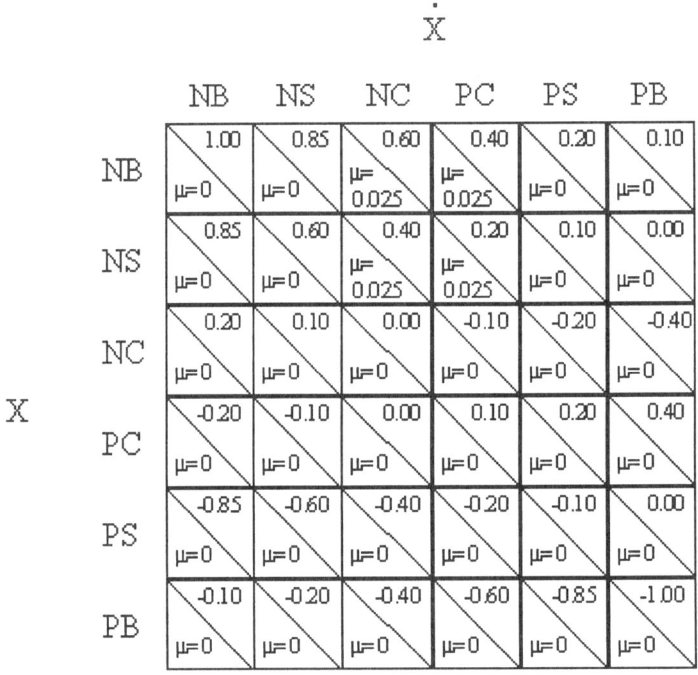

X \ Ẋ	NB	NS	NC	PC	PS	PB
NB	1.00 μ=0	0.85 μ=0	0.60 μ= 0.025	0.40 μ= 0.025	0.20 μ=0	0.10 μ=0
NS	0.85 μ=0	0.60 μ=0	0.40 μ= 0.025	0.20 μ= 0.025	0.10 μ=0	0.00 μ=0
NC	0.20 μ=0	0.10 μ=0	0.00 μ=0	-0.10 μ=0	-0.20 μ=0	-0.40 μ=0
PC	-0.20 μ=0	-0.10 μ=0	0.00 μ=0	0.10 μ=0	0.20 μ=0	0.40 μ=0
PS	-0.85 μ=0	-0.60 μ=0	-0.40 μ=0	-0.20 μ=0	-0.10 μ=0	0.00 μ=0
PB	-0.10 μ=0	-0.20 μ=0	-0.40 μ=0	-0.60 μ=0	-0.85 μ=0	-1.00 μ=0

Figure 4.4 - The abbreviated rule set with the corresponding weighting factors that are the product of the associated membership function values.

Now, the rule set must be applied in its entirety. Each rule recommends a potential action, and the recommended actions are not necessarily the same from one rule to another. Thus, the conflict provided by the rules must be reconciled so that a single crisp value of F_x is computed. This procedure has, to this point, been presented as a graphical calculation in which a COA is computed. However, this is not really necessary in the real-valued consequent approach. Actually, the weighted average of the applicable rules is taken — each rule is weighted with the product of the μ values associated with the condition portion of the rule. (If the minimum operator has been employed, then the rule would be weighted with the minimum μ value associated with the condition portion of the rule.) The mathematical formula is:

$$a = \sum_{i=1}^{n} \frac{a_i \, \mu_i}{\mu_i}$$

where a_i represents the real-value of the action in the ith rule and n is the number of rules. The fuzzy controller prescribes a value of 0.40 for the thrust in the x direction for the conditions given. Similar calculations are performed to compute values for the thrusts in the y and z directions.

A computer program was written to implement the fuzzy control system for the satellite rendezvous problem for a specific set of initial conditions. Figure 4.5 is a plot of the relative position of the spacecraft and the target. Notice that after approximately 600 seconds the rendezvous is complete. These results demonstrate that controllers that solve difficult, real-world control problems can be developed using the step-by-step procedure presented in Chapter 2. By the way, it is important to note that the authors are not experts in the art of spacecraft rendezvous. Thus, difficult problems can be solved by people who have only a crude feel for the system being considered.

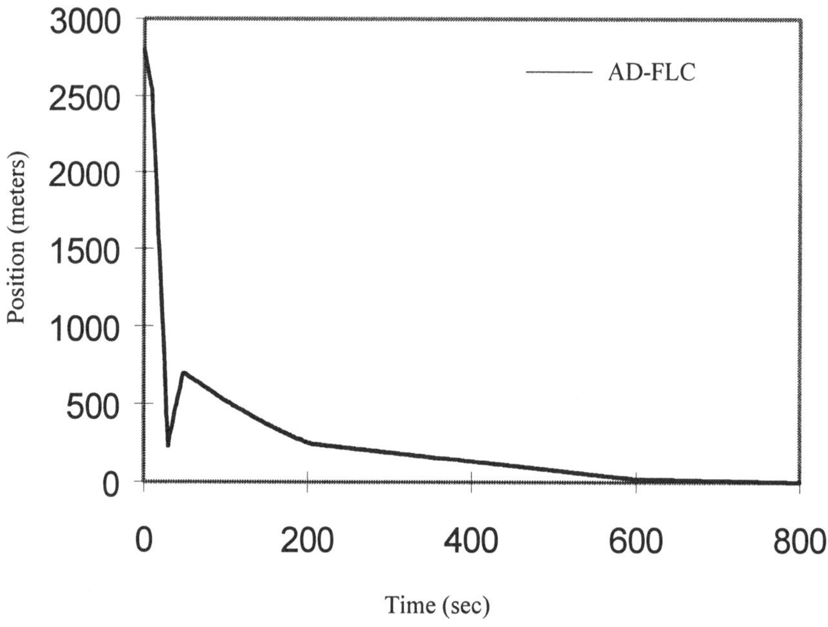

Figure 4.5 - The FLC successfully completes the rendezvous using rules and membership functions developed by the authors.

4.5 REVIEW/PREVIEW

In this chapter we have applied the step-by-step procedure set forth in Chapter 2 to a spacecraft rendezvous problem. We have used this procedure with the help of an expert to develop a controller for a difficult problem about which we actually knew little. In developing the rendezvous controller, we have introduced three new ideas.

The first new idea involves the defuzzification of the rules. In previous chapters, the minimum operator was used in which the final action set forth by the controller was determined using the minimum confidence associated with individual rules. In the current chapter, the multiplication operator is used in which individual confidences associated with a rule are multiplied together. Basically, this approach did not provide any better results (Karr, Fleming, and Vann, 1994); it is just an alternative method the reader should be familiar with since it is used in some fuzzy control systems.

The second new idea was the real-valued consequent approach. In this approach the rules have their actions described by real values instead of by linguistic terms. This rule form provides the user with greater flexibility in choosing the rule set, and allows for control systems that are very precise.

The third new idea involves the use of two different rule sets. In previous chapters, all of the rules had identical forms. However, because of the physics of the rendezvous system, it was necessary to write two types of rules. Specifically, rules were written for each of the coordinate directions individually, then the coupling of the system was addressed using a set of specific coupling rules. In adopting this approach, the system required only 124 rules instead of the 66656 rules required without this approach. We recommend exploiting information about the problem when possible.

The most noteworthy feature introduced in this chapter is the real-valued consequent approach. When we discovered this approach during the course of the rendezvous project, we were fascinated at the possibilities now available to us. However, we were faced with a major dilemma: how were we to select the floating point values appearing on the right-hand side of the rules? In fact, the rules for the rendezvous were extremely difficult to write even using the "old approach" in which fuzzy terms were used for the actions. We went through a difficult trial-and-error process in which we would write some rules, test the controller, refine the rules, test the controller, etc. Incidentally, this is the procedure that developers of fuzzy systems use routinely to get their systems to achieve the goals set forth for their controllers. Although the basic mechanics of the procedure are relatively straightforward, defining all of the rules and membership functions so that they work well in concert with one another is not an easy task.

The problem of defining the rules is very much related to the problem of defining the membership functions. There are an unlimited number of choices for both the membership functions and the rules, and they must work well together and neither set should be too large. Well, it is true that both of these ideas came to the forefront one day in our little research lab via a very intense philosophical discussion. The solution to the problem represents the authors' major contribution to the field of fuzzy logic, and also represents the end of Part I of this book. The next part of this book is the beginning of our journey into adaptive control; it is about the genetic algorithm and its role in rule and membership function selection. Using the GA we can produce a relatively small, co-adapted set of rules and membership functions.

REFERENCES

Burden, R. L., Faires, J. D., and Reynolds, A. C. (1978). *Numerical analysis.* Boston, MA: Prindle, Weber, & Schmidt.

Clohessy, W. H., and Wiltshire, R. S. (1960). Terminal guidance system for satellite rendezvous. *Journal of the Aerospace Sciences*, **Sept**, 653-674.

Kaplan, M. H. (1976). *Modern spacecraft dynamics & control.* New York: John Wiley & Sons.

Natenbruk, P., and Ragnittk, D. (1983). Control aspects as elaborated in space rendezvous simulations. *Conference Proceedings on Guidance and Control Techniques for Advanced Space Vehicles*, Florence, Italy.

PART II
GENETIC ALGORITHMS

Chapter 5

Genetic Algorithms as Search Procedures

5.1 INTRODUCTION

In Part I of this book, we introduced a procedure for developing fuzzy control systems. As the problems we addressed grew in complexity, the tasks of writing the rules and defining the fuzzy terms with membership functions became difficult and tedious. Now we present a search technique that reduces or eliminates the difficulty and tedium associated with selecting rules and membership functions.

Genetic algorithms (GAs) are our search technique of choice for developing adaptive control systems. There are a number of reasons that GAs are effective tools for developing adaptive control systems, and we will soon demonstrate many of these reasons. In this chapter, we introduce GAs and use them to solve least-squares curve-fitting problems. In a curve-fitting problem the parameters of an equation are given values so that the equation represents a set of data as closely as possibly.

Curve-fitting problems are of interest to us in our quest for adaptive control systems because they introduce us to methods that are useful for developing computer models. Computer models are needed to train the adaptive controllers. In fact, the curve-fitting problems we address in this chapter serve as an introduction to a very general method for adjusting parameters in complex empirical computer models — a task we will face numerous times before this book is complete, and a task that is discussed in much greater detail later in Chapter 10. However, the conventional least-squares method of curve-fitting has limitations; nonlinear forms and forms for which no derivative information exists present problems. These limitations can be bypassed using a GA.

GAs are search algorithms based on the mechanics of biological genetics (Holland, 1975). GAs generally operate on populations of binary strings that are coded to represent proposed solutions to a problem. They efficiently exploit knowledge contained in the population of strings, and generate strings that represent new and improved solutions to the problem of interest. GAs are used in a variety of problems (De Jong, 1975; Goldberg, 1983; Goldberg and Lingle, 1985; Goldberg and Samtani, 1986; Karr and Goldberg, 1987) where they are shown to rapidly locate near-optimal solutions.

In this chapter, a simple GA is applied to three least-squares curve-fitting problems. Although the problems can be solved effectively using more-conventional techniques, they serve to demonstrate the use of GAs In fact, in the two real-world examples presented, the GA is able to find better solutions than those found using alternative methods. Furthermore, the method for fitting curves to data using a GA is easily extended to more-complex problems that are difficult for the conventional least-squares technique. This is demonstrated in Chapter 10 where the approach is extended to the method of least median squares, which is a curve-fitting technique that is insensitive to outliers (errors or mistakes) in the data.

5.2 A SIMPLE GENETIC ALGORITHM

John Holland, the father of GA research, developed a mathematical theory that describes why and how GAs do what they do — a theory that involves a somewhat esoteric entity called *schemata* and the k-armed bandits (Holland, 1975). This mathematical theory is certainly a necessary component in a technology's voyage from laboratory novelty to industrial tool. However, in keeping with the policy we have set for this book, we choose to focus on how GAs work as opposed to why GAs work. There are at least two high quality texts on GAs from which readers learn the finer points of GA theory (Goldberg, 1989; Davis, 1991). The approach we use is to introduce and solve a specific problem using GAs, and then briefly describe why they work.

Consider the following curve-fitting problem: develop the equation of a line that provides the "best" fit to the following data:

x	f(x)
0.0	0.0
1.0	1.0
2.0	2.5
3.0	3.0
4.0	4.0
5.0	5.0

Notice that the data points fall exactly on the line described by $f(x)=x$, except for the point at $x=2.0$. This problem is easily solved using a wide variety of curve-fitting techniques. It is used here simply to introduce the basic steps required to implement a genetic algorithm. Later in this chapter some more-difficult curve-fitting problems are presented.

5.2.1 Getting Started

The first issue to address is to determine what we mean by the term "best." There are numerous curve-fitting strategies that can be used, but we will employ the conventional least-squares criteria. Thus, the problem can actually be solved with a calculus-based approach in which an error term is minimized. This approach provides the values m=1.057 and b=-0.0592 in the equation f(x) = mx + b. But, as will be seen later, the conventional calculus-based least-squares approach faces some challenging hurdles in more complex problems. Therefore, let's start expanding our curve-fitting tools; let's do something different. Let's solve the above problem using GAs, and then discuss why GAs work.

The first step in the implementation of a GA is to randomly generate a population of N binary bit-strings of length p. There are some accepted techniques used for determining the size of the population, N, and the length of the strings, p, but for now we will use N=5 and p=8. These values are good for demonstrating the operation of a GA, but not very good if we are actually trying to solve the problem accurately. Table 5.1 depicts the initial generation of five strings of length eight that were randomly generated.

string number	string
1	00111001
2	11011110
3	11100101
4	10000110
5	01000110

Table 5.1 - The randomly generated initial population of strings for a GA serves as a starting point in the least-squares problem.

5.2.1 Coding

The second step is to give meaning to the strings. Each bit string must represent a complete solution to the problem; our problem is to select values for two parameters, m and b, so that the equation f(x)=mx + b best fits the data. Thus, each string must be coded so that it represents both parameters. We will use a method called concatenated, binary, linear mapping in which a portion of the bits in each string represent the first parameter, m, and the remaining bits represent the second parameter, b. We will simply divide the bit strings equally so that the first four bits in each string represent m and the last four bits represent b. We will interpret each four-bit substring as a binary number. For those of you who have forgotten your binary, or never actually learned it, Table 5.2 contains the binary to base 10 conversions.

binary	base 10	binary	base 10
0000	0	1001	9
0001	1	1010	10
0010	2	1011	11
0011	3	1100	12
0100	4	1101	13
0101	5	1110	14
0110	6	1111	15
0111	7		

Table 5.2 - Binary to base 10 conversions are used in many GA encoding schemes.

Thus, using Table 5.2, we can convert a substring from a bit representation into a base 10 value. However, if that is all we do, then each parameter would have an integer value between 0 and 15. It is highly unlikely that our values of m and b are well represented by an integer. In fact, we know for this problem that our precise values are m=1.057 and b=-0.059. Thus, we need to map the values from binary to decimal within a restricted and different range. We can look at the data in our particular problem and see that the values of m and b are probably going to be somewhere between m_{min} = 0.5, m_{max} = 1.5, b_{min} = -0.5, and b_{max} = 0.5. There is nothing in the GA theory that helps the user select these values; they are dependent solely on the particular problem being solved and must be supplied by the user. Once, however, these values are selected, the binary strings can be mapped according to the formula:

$$m = m_{min} + \frac{bin}{(2^p - 1)}(m_{max} - m_{min})$$

where bin is the binary value of the substring and p is the length of the substring. Thus, with the minimum and maximum values provided, the value of m represented by string number 1 in Figure 5.1 (0011) is

$$m = 0.5 + \frac{3}{(2^4 - 1)}(1.5 - 0.5) = 0.7$$

The value of b represented by string number 1 is

$$b = -0.5 + \frac{9}{(2^4 - 1)}(0.5 + 0.5) = 0.1$$

Thus, string number 1 in our initial population represents a potential solution of m=0.7 and b=0.1. Each parameter, m and b, represented by different binary numbers all have a value that is mapped linearly between the minimum and maximum values selected by the user. Table 5.3 shows the values of the parameters associated with the possible substrings in the coding as it has been defined and demonstrated in Table 5.1. Furthermore, Table 5.4 is an expanded version of Table

5.2 in which the parameter values associated with the strings in the initial population have now been included.

binary	base 10	m-value	b-value
0000	0	0.500	-0.500
0001	1	0.567	-0.433
0010	2	0.634	-0.366
0011	3	0.701	-0.300
0100	4	0.767	-0.234
0101	5	0.834	-0.167
0110	6	0.900	-0.100
0111	7	0.967	-0.034
1000	8	1.034	0.032
1001	9	1.101	0.099
1010	10	1.167	0.166
1011	11	1.233	0.232
1100	12	1.300	0.300
1101	13	1.367	0.366
1110	14	1.433	0.433
1111	15	1.500	0.500

Table 5.3 - The parameter values are mapped linearly between $m_{min} = 0.5$, $m_{max} = 1.5$, $b_{min} = -0.5$, and $b_{max} = 0.5$.

string number	string	m	b
1	00111001	0.701	0.099
2	11011110	1.367	0.433
3	11100101	1.433	-0.167
4	10000110	1.034	-0.100
5	01000110	0.767	-0.100

Table 5.4 - The initial population of strings for a GA in the least-squares problem with the parameter values obtained using the particular coding described above.

5.2.3 Fitness Function

Now that each string has some "meaning" in our particular curve-fitting problem, the GA requires us to provide a value of relative merit for each possible solution; a fitness function that informs the GA of how close it is to achieving our objective. In this problem, our objective is to minimize the sum of the residuals between the data points and our particular curve. Mathematically, we want to minimize the following

$$v = \sum_{n=1}^{n=M} (y_i - y_{i'})^2$$

where v is the value to be minimized, y_i is the value of f(x) associated with data point i, y_I' is the corresponding value of f(x) predicted by the model equation, $y_I' = mx_i + b$, and M is the number of data points.

This value v, however, does not correspond directly to the fitness value needed by a GA. Although GA operators have been developed that can be used for minimization problems, the operators used in this chapter (to be introduced shortly) allow only for solving maximization problems. Thus, we must convert our minimization problem into a maximization problem. To do this, we will simply subtract the value of v from a large number. Thus, an acceptable fitness value, f, is as follows

$$f = K - \sum_{n=1}^{n=M} (y_i - y_{i'})^2$$

where K is a constant just large enough to make all of the fitness values greater than zero. In the current example, K = 25.0. This function requires good solutions to have fitness values that are larger than those of poor solutions. With this in mind, Table 5.5 shows the fitness values associated with each of the strings in the initial generation.

string number	string	m	b	fitness
1	00111001	0.701	0.099	20.17
2	11011110	1.367	0.433	12.61
3	11100101	1.433	-0.167	17.14
4	10000110	1.034	-0.100	24.70
5	01000110	0.767	-0.100	20.44

Table 5.5 - The initial population of strings for a GA in the least-squares problem with the parameter and fitness values indicates that string 4 is the best performer.

5.2.4 Generate New Solutions

At this point, we have obtained five random possible solutions to the least-squares problem. It is time to work some GA magic through which the information contained in this initial generation will produce subsequent generations that contain better solutions to the problem.

The next step in the process is to apply the GA operators. Although numerous high-powered, problem-specific GA operators have been developed and used successfully, most problems can be solved very effectively using only three basic operators: (1) reproduction, (2) crossover, and (3) mutation.

The first operation, reproduction, allows strings with large fitness values, good solutions to the problem at hand, to receive many copies in the new population. In this chapter, use is made of a particular type of selective reproduction, called expected number control. This form of selective reproduction makes a certain number of copies, num_i, of a string in accordance with the equation:

$$num_i = \frac{f_i}{f}$$

where num_i = the number of copies of the string in the next generation,
f_i = the fitness of the individual string in the current generation,
and f = the average fitness of the current generation.

Thus, reproduction is the survival-of-the-fittest aspect of the GA. The best strings receive more copies in subsequent generations so that their desirable traits can be passed to their offspring. It should be apparent that in this form of reproduction, high quality strings must have large values of fitness because the GA is going to maximize the fitness values.

The newly formed strings are placed in a file, generally called a mating pool, where they await application of the second and third operators. Table 5.6 depicts the mating pool resulting from the application of the reproduction operator to the initial generation. Notice that the mating pool does not contain any strings that were not represented in the initial generation, only copies of the selected strings. Rather, the reproduction process has simply exploited the fitness function to bias the new population toward high-quality strings. Thus, the remaining operators must produce strings containing new and better parameter values if we ever hope to get significantly better solutions than those represented by the initial, randomly generated population. The remaining operators must generate parameter values that have not yet been considered; they must investigate new points in the search space.

string number	string	m	b	fitness
1	00111001	0.701	0.099	20.17
2	10000110	1.034	-0.100	24.70
3	01000110	0.767	-0.100	20.44
4	10000110	1.034	-0.100	24.70
5	01000110	0.767	-0.100	20.44

Table 5.6 - The mating pool for the least-squares problem contains relatively large numbers of highly fit strings.

The second operator, crossover, affords a means for strings to mix and match their desirable qualities through a random process. After reproduction, simple crossover proceeds in three steps. First, two strings are selected from the mating pool. Second, a position along the strings is selected at random. This is shown as follows, where two binary coded strings X and Y (strings number 1 and 2 in the mating pool) are selected for crossover:

X = 0 0 1 | 1 1 0 0 1
Y = 1 0 0 | 0 0 1 1 0

Crossing site 3 has been selected in this particular example, although any of the other six positions could have been selected. The third step is to exchange all characters following the crossing site. The two new strings produced by this crossover are shown as X' and Y':

X' = 0 0 1 0 0 1 1 0
Y' = 1 0 0 1 1 0 0 1.

String X' is made up of the first part of string X and the tail of string Y. Likewise, string Y' is made up of the first part of string Y and the tail of string X. Crossover is an effective means of exchanging information and combining portions of high-quality solutions to obtain even better solutions. Although crossover has a random element, this operator does not result in a random search. Table 5.7 depicts the population after crossover has been applied.

string number	string	m	b	fitness
1	10011001	1.101	0.099	24.13
2	00100110	0.634	-0.100	15.39
3	01000110	0.767	-0.100	20.44
4	01000110	0.767	-0.100	20.44
5	10000110	1.034	-0.100	24.70

Table 5.7 - The reproduction and crossover operators result in a generation of strings that have a higher maximum fitness value.

The mechanics of the reproduction and crossover operations are quite simple; they involve nothing more than making copies of strings and exchanging portions of strings. However, reproduction and crossover together give GAs most of their power.

The third operator, mutation, protects the ability of the GA to find near-optimal solutions. Mutation is the occasional alteration of a bit at a randomly selected position in a string. It insures against the loss of a particular piece of information. Reproduction and crossover may together create a generation that is void of a particular character at a given string position. For example, a generation may arise that does not have a 1 in the third string position. Neither reproduction nor crossover will ever produce a 1 in this third position in subsequent generations. The mutation operator simply flips bits from 1 to 0, or 0 to 1 at random. Thus, mutation allows a 0 in the third position to be changed to a 1, thereby reinstating the critical piece of information into the population. Although mutation is a vital part of any GA, it occurs with a small probability (on the order of one mutation per thousand string positions). Since mutation is performed with such a small probability, in our example in which there are 5*8=40 bit positions, the random forces driving mutation did not change any bits in the first generation, so that the second generation is the same as that shown in Table 5.7 above.

At this point, we have developed a second generation of strings that contains better solutions to the search problem than existed in our initial, randomly generated population. We now go back to the beginning and decode our strings, evaluate their fitnesses, and apply the three operators again. We continue with this process of generating new populations until we decide to stop (and there are numerous stopping criteria including reaching a maximum number of generations, finding an acceptable fitness value, and running until some environmental factor dictates that we stop).

Figure 5.1 shows graphically the best solution found by the GA at selected generations to give the reader an indication of how this process proceeds through the search space. Each figure includes the optimum solution for comparison purposes. Notice that after only 10 generations, the GA has found a solution that is very close to the optimum. Additionally, Figure 5.2 shows the strings present in various generations. The values associated with the strings in various generations are plotted on an m-b axis system. The plotted points in a particular generation seem to be converging on the theoretical best solutions. The figures show the sampled points in the search space at selected generations. Notice that even though the GA continues to sample points throughout the search space defined by the parameters b and m, most points in later generations cluster about the best solution.

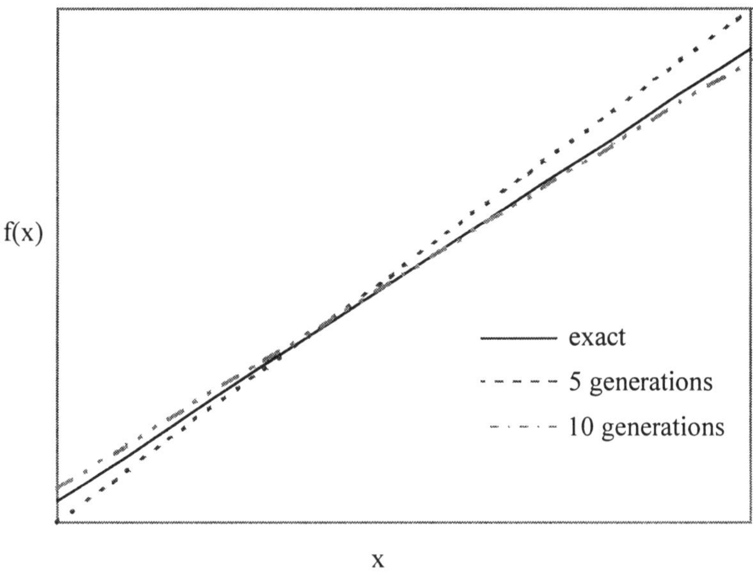

Figure 5.1 - As the GA considers more and more alternatives, its "proposed" solutions move closer and closer to the optimum.

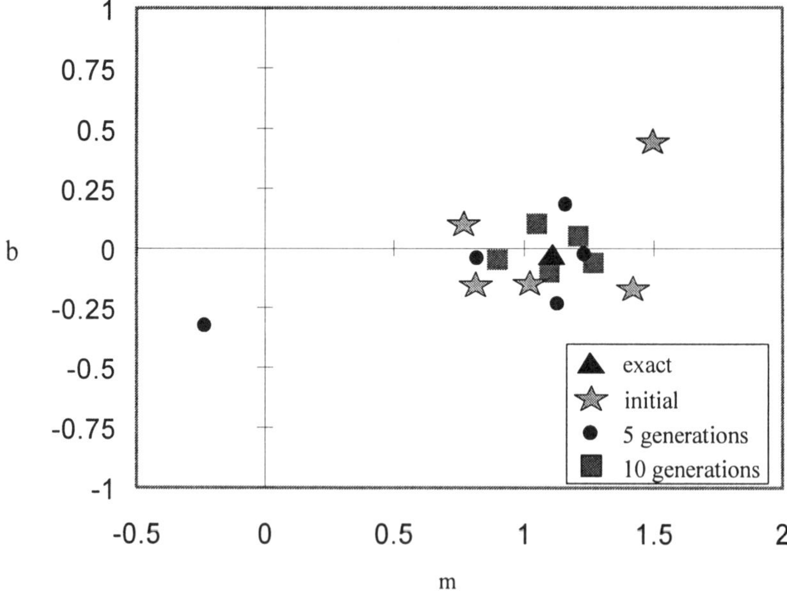

Figure 5.2 - Notice that the population of points seem to converge to the correct solutions as the number of generations increases.

This has been a brief overview of a simple GA. For further details of the processing power and the convergence properties of GAs, refer to Goldberg (1989) or Davis (1991).

5.3 HOW GENETIC ALGORITHMS WORK

Effective search algorithms must, in some way, exploit information about the search space. Derivative-based search schemes utilize information about gradients to direct their walk through the search space. This exploitation of derivative information often leads to an effective search algorithm. However, there are problems in which derivative information is misleading, and thus derivative searches fail. GAs tend to exploit information in a radically different way.

GAs work with binary representations of the parameters. Each binary string represents a single point in the search space. Each population of strings (each generation) represents a cloud of points in the search space. The GA operators of reproduction, crossover, and mutation serve to reshape the cloud of points from one generation to the next and to concentrate it in the region of the solution. No derivative information is used. The only information a GA exploits is information contained in the fitness values. The fitness values are used by the GA to find regions of the search space that contain highly fit points (good solutions). This information is used by the reproduction operator to bias the cloud of points in a generation toward or into these regions. Thus, the fitness information is exploited and guides the search.

GAs effectively balance the two key aspects of any search: (1) exploitation and (2) exploration. Exploitation uses local information to refine local solutions, while exploration is a way of insuring that global objectives are met; here, global is used to reference the entire search space. Effective search techniques must balance the concepts of exploitation and exploration. Derivative-based searches tend to rely exclusively on exploitation. They have no mechanism for recognizing when their search has become trapped in a local optimum. Monte-Carlo searches, on the other hand, tend to rely exclusively on exploration and totally neglect exploitation; Monte-Carlo techniques are basically random walks through a space. GAs actually balance these two ideas quite well. Reproduction exploits fitness information to bias the points being considered toward high-fitness regions. Crossover and mutation, on the other hand, do not rely on fitness information. These two operators place randomly generated points in the population. Crossover and mutation allow for the consideration of points far from the current local optimum.

GAs exhibit some fundamental differences from more-conventional search techniques, most of which are calculus-based. These differences include:

1. GAs work with bit strings that represent entire parameter sets, whereas most calculus-based methods manipulate individual parameters independently.

2. GAs work with numerous points simultaneously through their manipulation of populations. Thus, they are better able to balance the ideas of exploitation and exploration.

3. GAs use random choice as part of their search strategy (specifically in both the crossover and reproduction operators).

5.4 GENETIC ALGORITHM APPLIED TO LEAST-SQUARES CURVE-FITTING PROBLEMS

In this section, a simple GA is used to solve two additional least-squares curve-fitting problems. The purpose of including these additional examples is to instill confidence in the reader that a genetic algorithm can be used to solve curve-fitting problems that are more difficult than the simple line presented in section 5.2. In both problems, the details of the genetic algorithm's application are consistent with the example of section 5.2. Thus, only the results are presented. In both problems the GA finds a solution that is better than previously found using alternative techniques.

5.4.1 Preliminary Considerations

Before considering the two specific problems, two points concerning the application of the GA must be addressed. There are basically two decisions to be made when applying a GA to any given problem: (1) how to code the parameters of the problem as a finite string, and (2) how to evaluate the merit of each string (each parameter set).

The method for coding the strings discussed previously will again be used here. A linearly mapped, binary, concatenated coding scheme is used. In this scheme, the search parameters are represented as binary strings. Substrings representing the individual parameters are concatenated to form the full strings. The binary representations are mapped linearly between minimum and maximum parameter values.

Consider the question of how to evaluate each string. In least-squares curve-fitting problems, the objective is to minimize the sum of the squares of the distances between a curve of a given form and the data points. Thus, if y is the ordinate of one of the data points, and y' is the ordinate of the corresponding point on the curve produced by the modeling equation, the objective of least-squares curve fitting is to make $(y - y')^2$ a minimum. This square of the error, which is to be minimized, affords a good measure of the quality of the solution. However, the GA seeks to maximize the fitness. Again, the minimization problem must be transformed into a maximization problem. Here, again, the error is simply subtracted from a large positive constant (K) so that

$$f = K - (y - y')^2$$

yields a maximization problem.

The selection of an appropriate coding and the determination of a fitness representation are the only aspects of a GA that are problem specific. Once these two decisions have been made, every problem is the same to a GA. The GA generates populations of strings, evaluates them, and uses these evaluations to produce subsequent generations of more highly fit strings. Thus, the GA is a very flexible search method.

The simple GA described earlier has been programmed in the C programming language by modifying the code found in Goldberg (1989). The GA is run in these examples with the following parameters:

Number of strings in the population (popsize) = 100
Probability of crossover ($P_{crossover}$) = 0.65
Probability of mutation ($P_{mutation}$) = 1 / popsize = 0.01

These values are consistent with De Jong's suggestions (1975) for moderate population size, high crossover probability, and low mutation probability.

5.4.2 Example 1 — Polymer Required for Dewatering Versus Concentration

The U.S. Bureau of Mines investigated a method of dewatering flocculated fine-particle waste that involves flocculating clay with a high-molecular-weight polymer such as polyethylene oxide (PEO). Stanley and Scheiner (1985) found an empirical relationship between the amount of polymer required for dewatering clay and the polymer concentration. The equation is of the form

$$A^{C_1} = C_2 C + C_3$$

where

A = the amount of polymer (PEO) required in gram per kilogram of clay to dewater a particular clay (the dependent variable),

C = concentration, in weight percent, of the polymer solution used for dewatering (the independent variable),

C_1 = constant that determines the curvature of the plot, and

C_2, C_3 = constants defining the line.

This problem is difficult to solve using a traditional least-squares approach. Thus, to establish the correlation, A is raised to the C_1 power and plotted versus C. This method is illustrated for an unexchanged attapulgite clay dewatered with 0.05 pct PEO solution. The data points are plotted for values of C_1 of 1.7, 2.1, and 2.5 as shown in Figure 5.3. A unique straight line is obtained at 2.1. This selection method is a trial-and-error process. The use of a GA in fitting data to an equation of the given form shows its effectiveness in solving non-linear curve-fitting problems.

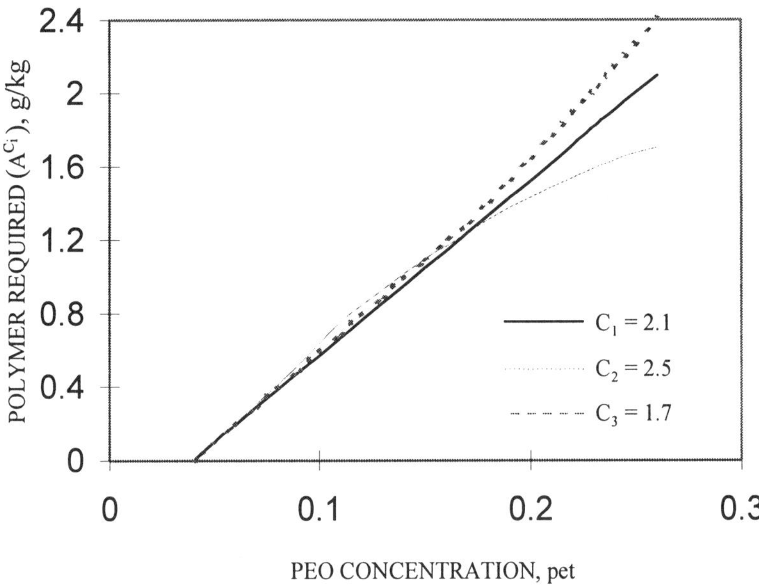

Figure 5.3 - Trial-and-error process for determining the coefficients associated with the parameters in the dewatering equation.

The addition of a third constant makes this example more difficult than the line represented by data points considered in the previous section because the size of the search space has increased significantly (from 2^{22} to 2^{33} possible solutions). Parameter sets are coded in the same manner as those of the line, except the strings representing the parameters now have a length of 33; the first 11 bits represent C_1, the second 11 bits represent C_2, and the final 11 bits represent C_3. The constants are once again determined based on the minimization of the square of the distance between the points and the curve. C_1, C_2, and C_3 were considered to range from -32.0 to 32.0.

Three data points were fitted for both aluminum-exchanged and potassium-exchanged clays. Figure 5.4 shows the best-of-generation error for two independent runs as it decreases with successive generations. Again the results of the GA quickly converge to near-optimal solutions. The results produced by the GA are compared with those of Stanley and Scheiner's (1985) trial-and-error method in Figure 5.5. The constants produced by the GA are compared with those of Stanley and Scheiner in Table 5.7.

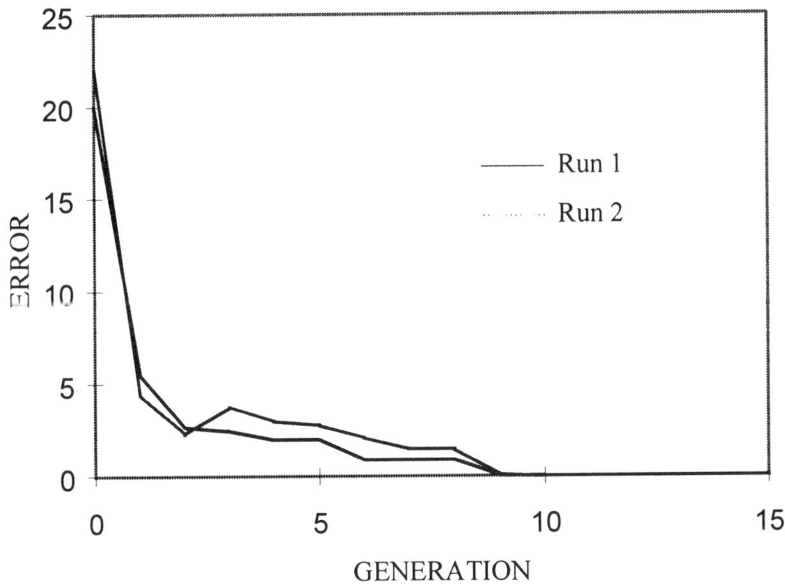

Figure 5.4 - Best-of-generation error versus generation number for example 1.

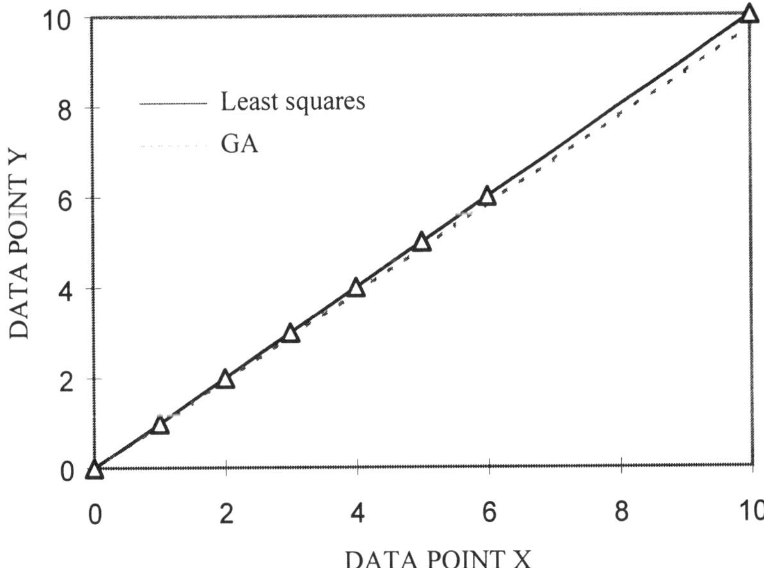

Figure 5.5 - GA-predicted curve fit for example 1.

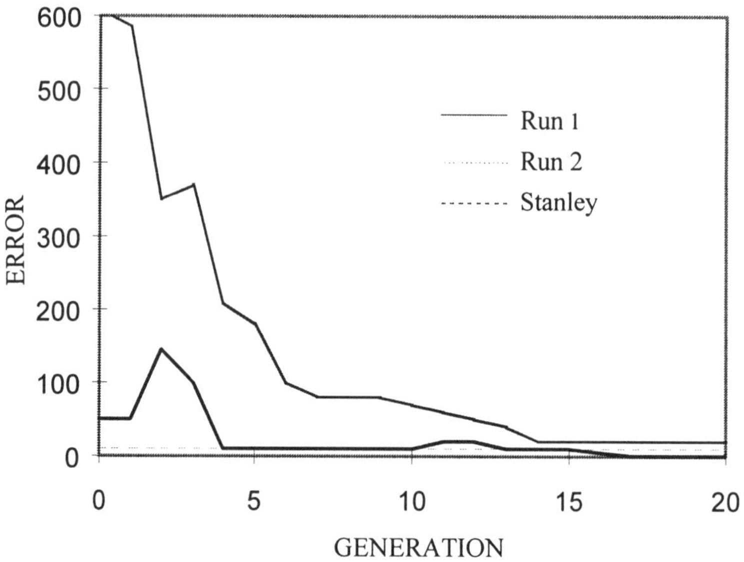

Table 5.6 - Comparison of trial-and-error and GA constants.

	K^+	Al^{3+}
Trial-and-error:		
C1	1.40	2.00
C2	16.39	7.20
C3	0.0	0.0
Genetic Algorithm:		
C1	1.40	2.00
C2	16.280	7.063
C3	-0.031	0.031

Table 5.7 - Comparison of coefficient values found using the trial-and-error method and those found using a genetic algorithm.

5.4.4 Example 2 — Ree-Eyring Equation

The Ree-Eyring equation allows for the estimation of the Newtonian and non-Newtonian components of viscosity and is of interest to the mineral processing industry. The Ree-Eyring equation can be written as

$$y = C_1 + C_2 \frac{\sinh^{-1}(C_3 x)}{C_3 x)}$$

where

y	= the viscosity (the dependent variable),
x	= the shear rate (the independent variable),
and	
C_1, C_2, C_3	= the constants to be determined.

The process of fitting the data using a GA is the same as in the previous example. The parameter set (C_1, C_2, and C_3) is again represented as a 33-bit string of zeros and ones and is determined by minimizing the square of the distance between the data points and the curve.

Since this equation cannot be fitted to data using a direct least-squares approach, Stanley, Webb, and Scheiner (1986) used a transformation method to solve the Ree-Eyring equation for a sodium ion-exchanged clay. The data used in that report were also used here, and the results of the curve produced using the GA are compared with those previously obtained. Once again the GA quickly converged an excellent solution. Figure 5.6 shows the best-of-generation error for two independent runs as it decreases with successive generations. The smallest error produced by the GA calculated curve (1.57) is less than that of the curve calculated by the transformation method used by Stanley, Webb, and Scheiner (12.55). Figure 5.7 shows how well the curve calculated by the GA (C_1=168.83, C_2=34.20, and C_3=0.043) fits the data. C_1, C_2, and C_3 ranged between -200.0 and 200.0.

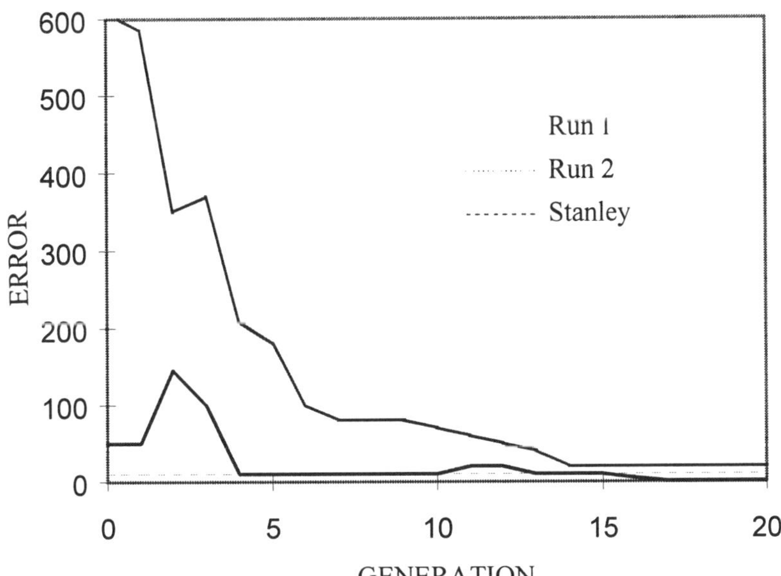

Figure 5.6 - Best-of-generation error versus generation number for example 2.

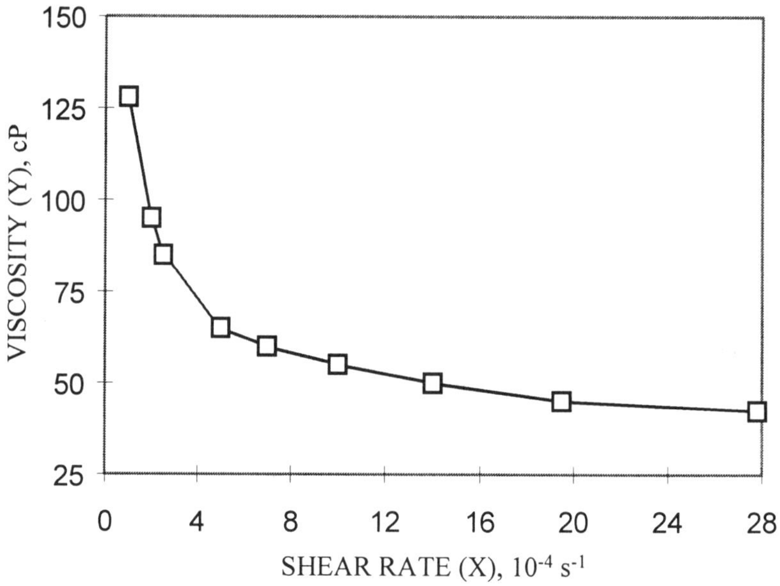

Figure 5.7 - GA-predicted curve fit for example 2.

5.5 REVIEW/PREVIEW

In this chapter, a simple GA made up of reproduction, crossover, and mutation was used to perform least-squares curve-fitting. Three curve-fitting problems were presented in which a GA rapidly converged to near-optimal solutions. The solutions were comparable to, or better than, solutions determined by more commonly used methods. Although they require some knowledge of the problem, GAs are readily adaptable to least-squares curve-fitting problems.

Determining the solution to least-squares curve-fitting problems is a challenging and worthwhile endeavor. However, we are more interested in using a GA-based search technique to select effective rules and tune membership functions in the fuzzy control and modeling systems we wish to implement. It is not readily apparent at this point how we can use a GA to accomplish these tasks. There are several points that must be addressed for the reader to be able to design and tune a fuzzy system using a GA.

The next three chapters will address many of the relevant issues that arise when using GAs to design fuzzy systems. Chapter 6 revisits the liquid level system first introduced in Chapter 2. The fuzzy controller is improved with a GA. Chapters 7 and 8 revisit the cart-pole system and the rendezvous system, respectively. These three chapters provide information on the techniques that can be employed when using a GA to select fuzzy rules, to alter triangular and trapezoidal membership functions, and to determine real-valued consequents. Then, Chapter 9 will change the game dramatically when the systems to be controlled are made more complex.

REFERENCES

Davis, L. D. (1991) (Ed.), *The genetic algorithms handbook.* New York, NY: Van Nostrand Reinhold Company.

De Jong, K. A. (1975). Analysis of the behavior of a class of genetic adaptive systems. *Dissertation Abstracts International*, **36**(10), 5140B.

Goldberg, D. E. (1983). *Computer-aided gas pipeline operation using genetic algorithms and rule learning*, Ph.D. Thesis, University of Michigan, Ann Arbor, MI.

Goldberg, D. E. (1989). *Genetic algorithms in search, optimization, and machine learning.* Reading, MA: Addison-Wesley.

Goldberg, D. E., and Lingle, R. L. (1985). Alleles, loci, and the travelling salesman problem. Paper in *Proceedings of an International Conference on Genetic Algorithms and Their Applications*, J.J. Grefenstette (Ed.), 154-159.

Goldberg, D. E., and Samtani, M. P. (1986). Engineering optimization via genetic algorithm. Paper in *Proceedings of Ninth Conference on Electronic Computation*, University of Alabama in Birmingham, 471-482.

Holland, J. H. (1975). *Adaptation in natural and artificial systems.* Ann Arbor, MI: University of Michigan Press.

Karr, C. L., and Goldberg, D. E. (1987). Genetic algorithms in mineral processing and machine learning. Paper in *Proceedings of the Artificial Intelligence in Mineral and Material Technology Conference*, University of Alabama, Tuscaloosa, AL, 127-141.

Stanley, D. A., and Scheiner, B. J. (1985). Mechanically induced dewatering of ion exchanged attapulgite flocculated by a high molecular weight polymer. *Colloids and Surfaces*, **14**, 151-159.

Stanley, D. A., Webb, S. W., and Scheiner, B. J. (1986). *Rheology of ion-exchanged montmorillonite clays.* Bureau of Mines RI 8895.

Chapter 6

Genetic Algorithm Design of a Liquid Level Controller

6.1 INTRODUCTION

Chapter 2 of this book presented a step-by-step procedure for developing fuzzy systems, and applied it to the design of a fuzzy controller for a simple liquid level system. The development was straightforward and produced a controller that worked well. However, there are three factors that dramatically affect performance and so must be carefully adjusted by the developer: (1) the shape of the membership functions, (2) the location of the membership functions, and (3) the action values that characterize each linguistic rule. The shape and location of the membership functions determine the nature of the interpolation between anchor points (see Chapter 1). The action values of the rules locate the anchor points. The area of fuzzy system tuning has been under-investigated.

At the beginning of the project, a thorough literature search was conducted to discover published results involving fuzzy system tuning. This search, and conversations with other fuzzy system developers failed to reveal any accepted, or even any efficient algorithms for defining membership functions. At best the selection of membership functions was a difficult, time-consuming task; at worst, it was a task entwined with other facets of the fuzzy system such as the choice of implication operators (discussed in Chapter 8). Thus, since we had experience in using GAs for equipment design (Karr, 1989), GAs were a natural tool for us to use to define membership functions. It did not take the authors long to determine that GAs are effective tools for fuzzy membership function design.

Chapter 5 introduced genetic algorithms as search routines. The current chapter is an extension of Chapter 5 in the sense that it extends the use of GAs to adjust the parameters that characterize membership functions. Chapter 7 will introduce the use of GAs for selecting rule sets. These two ideas represent the first major contribution the authors have made to the field of fuzzy system development. The use of GAs solved a rather difficult and time-consuming aspect of fuzzy system development. The difficulty is the tuning of a fuzzy system so that it efficiently solves a particular control problem. Tuning a fuzzy system refers to the definition of the fuzzy membership functions and to the formation of an efficient rule set. Both the membership functions and the rule set must be adjusted simultaneously to obtain a well tuned and efficient fuzzy control system.

Our decision to employ GAs for defining membership functions is a decision we have not regretted. Most of the fuzzy systems we have developed over the years, and all of the fuzzy systems appearing in the remainder of this book, have been designed with the aid of a GA. In our experience, this approach is effective and saves time in the tuning of fuzzy systems.

This chapter provides the details necessary for employing a GA to improve the performance of the simple liquid level fuzzy controller presented in Chapter 2. The key issues of selecting a coding scheme and of defining a fitness function for this problem are addressed. Results are presented comparing the effectiveness of the author-designed (AD) liquid level controller with the GA-designed liquid level controller.

6.2 GENETIC ALGORITHM FOR MEMBERSHIP FUNCTION ADJUSTMENT

At the time this research effort began, the effectiveness of GAs in solving complex problems had been demonstrated in only a few engineering disciplines. However, it was understood from theory that the successful application of GAs to search problems required addressing two points: (1) how to represent (code) possible solutions to the search problem as a string of characters and (2) how to evaluate the effectiveness of possible solutions using a fitness function. Thus, consider these two points in relation to fuzzy membership function determination for the liquid level fuzzy controller.

6.2.1 Coding Scheme

The liquid level fuzzy controller presented in Chapter 2 employed triangular membership functions. Thus, the following presentation describes the use of a GA to adjust triangular membership functions. However, the same method can be applied to other membership function forms such as the trapezoidal membership functions used in the cart-pole controller of Chapter 3.

The unoptimized, arbitrarily selected, triangular membership functions used initially for the liquid level controller were shown in Chapter 2 (Figures 2.2 through 2.4). Here, the membership functions for Error (E) are repeated in Figure 6.1. Recall that there are four fuzzy sets used to describe error (E), five fuzzy sets used to describe the time rate of change of error (ΔE), and five fuzzy sets to describe the percent change in flow rate (Q_{net}). These fuzzy sets provide linguistic terms that can describe both the condition (E and ΔE) and the action (Q_{net})

variables. With these definitions, there are 20 rules in the controller. The membership functions determine under precisely what conditions, and to what extent each rule is applied for particular values of E and ΔE. Improving these membership functions improves the performance of the controller.

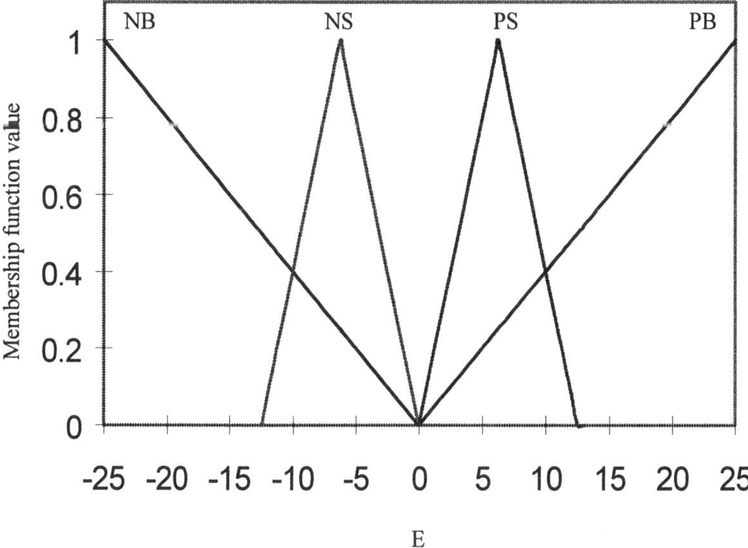

Figure 6.1 - The linguistic terms used to describe the condition and action variables for the liquid level controller are initially defined with the triangular membership functions shown above.

As a constraint on the GA's search, the membership functions in the liquid level controller were forced to maintain their triangular shape and their height, but the size of their base and their position relative to one another was allowed to change. The triangles describing ***extreme membership functions*** (those right triangles at the upper and lower limits of the variables they were describing; terms like **NB** and **PB** for E and ΔE; **BN** and **BP** for change in Q) were forced to remain fixed to their associated extreme limits. In other words, they were forced to remain right triangles with their apex at either the lower or upper limit of the condition or action variable they describe. Membership functions defining ***interior membership functions*** were forced to remain isosceles triangles. Thus, the definition of each extreme triangle requires a single point (a left or right base point), while the definition of each interior membership function requires two points (both a left and a right base point). Figure 6.2 shows the six points that must be defined under these constraints to completely define the membership functions for E. (This figure has membership functions that are slightly different from those shown in Figure 6.1 so that the six points can be clearly identified.) When the membership functions associated with

E, ΔE, and Q_{net} are considered, the result is a search problem with 22 parameters that must be defined (six points defining the triangles for E and eight points each defining the triangles for ΔE and Q_{net}).

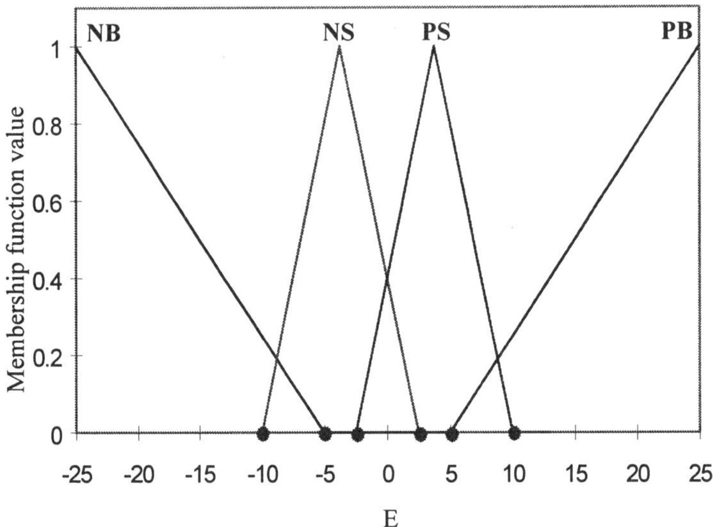

Figure 6.2 - The entire set of membership functions defining the linguistic terms used to describe the error in liquid level setpoint can be described with the definition of the six points identified above with a •.

Thus, the search for efficient membership functions for the liquid level control problem defined in Chapter 2 involves the selection of values for 22 parameters (22 points are needed to locate the 14 triangles that describe the fuzzy terms). This is a much larger search problem than the curve-fitting problems addressed in Chapter 5. However, again, the application of GAs to search problems requires only the coding of the parameters and the definition of a fitness function; the size of the problem does not prevent the use of a GA. It might, however, force an increase in either the population size or the maximum number of generations, or both.

In Chapter 5, the curve-fitting problems were solved with a GA using a concatenated, mapped, unsigned binary coding. In this coding, parameters that vary continuously between some minimum and maximum values are represented with bit strings. This same approach is applicable to the points defining the triangular membership functions in the current problem. For instance, a substring can be formed to represent the right base point of the extreme membership function **NB** describing E as shown in Figure 6.2. The substring is allocated a specified number of bits, p, and mapped using the familiar formula

$$C = C_{\min} + \frac{bin}{(2^p - 1)} * (C_{\max} - C_{\min})$$

where *bin* is the integer value represented by a *p*-bit string, C_{min} is the minimum value of the parameter, C_{max} is the maximum value of the parameter, and C is the value of the parameter represented by the substring. Each of the remaining parameters can be represented with similar substrings, and the entire parameter set can be represented by a string formed through the concatenation of the individual substrings. Again, the GA manipulates a population of strings using the familiar operators of reproduction, crossover, and mutation.

The remaining tasks include determining the number of bits to be allocated for each of the 22 parameters and the selection of appropriate minimum and maximum values for each of the parameters. To keep things simple, and because there is no apparent reason for using substrings of different lengths, each substring is represented with four bits. Thus, the resulting strings are thus 88 bits in length. In other problems it will be necessary to use substrings with more or fewer bits (i.e., different lengths) and the current coding scheme can certainly accommodate such a need.

Initially, the selection of minimum and maximum values for each of the parameters seems to be an obvious decision; it seems that the minimum and maximum parameter values should simply coincide with the lower and upper limits of a variable's range, i.e., the minimum value of the parameters for the membership functions describing E is -25 and the maximum value is 25. However, the selection of minimum and maximum values for each of the membership functions is at present subject to three constraints. The first constraint requires that each parameter value fall in the range selected by the user to represent each variable. The second constraint is that the definition of the linguistic terms should be consistent with the normal meaning of the terms. For example, the membership function associated with the term NEGATIVE BIG should be zero for positive numbers and near-zero for small negative numbers. The third constraint is that each point in the range being described should have at least one membership value assigned to it. In Figure 6.3 all the points on the E-axis have membership values defined by one or more of the membership functions except for those between -15.0 and -10.0. This is not acceptable. Membership functions must be constructed so that all relevant points have a membership value. Figure 6.4 illustrates a method for assuring that all points are associated with appropriate membership function values. Here, each point is allowed to slide within a pre-defined window bounded by values of C_{min} and C_{max}. The permitted ranges are selected by the user, but, as shown in the figure, these ranges must assure that there will be a small degree of overlap of the membership functions.

Actually, the three constraints listed above should be viewed as soft constraints. Generally speaking, they are solid guidelines the user should follow. However, there are situations in which these constraints can be violated, yet viable fuzzy systems can still be obtained. Later in this book, we will encounter such situations. For example, the hexamine controller we introduce in Chapter 15 employs a GA to continuously adjust membership functions in real-time. The hexamine controller will actually have gaps in portions of the state space in which the controller is not currently operating. However, since adjustments are being made continuously, the gaps are eliminated before the controller operates in that region of state space. For now, the three constraints will be followed for two reasons: (1) they are intuitive and (2) they will help keep us out of trouble.

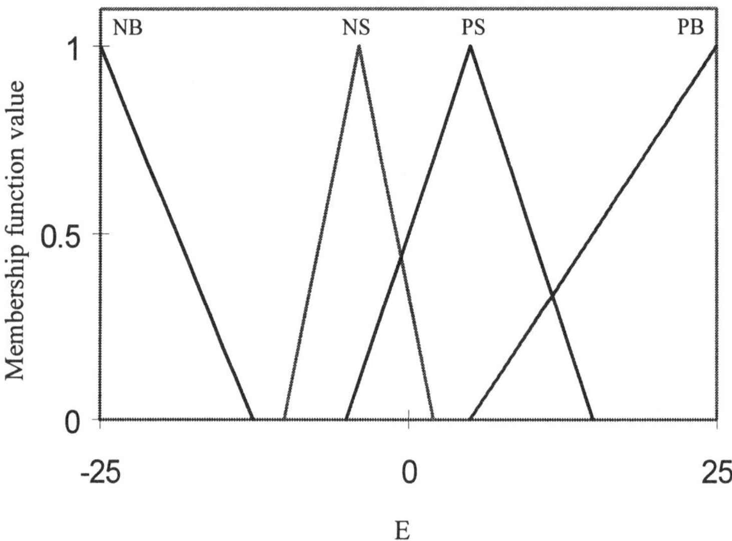

Figure 6.3 - If the values of C_{min} and C_{max} are not selected carefully, controllers with gaps in their condition variable ranges can result. These gaps are indicated by regions for which the fuzzy sets do not overlap.

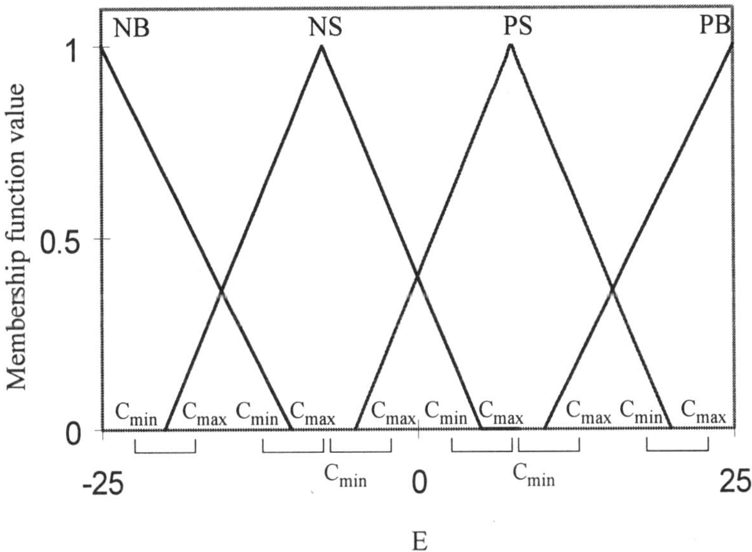

Figure 6.4 - The values of C_{min} and C_{max} shown above eliminated the possibility of a GA considering fuzzy systems with gaps in them.

6.2.2 Fitness Function

The fitness function employed by a GA used to locate efficient fuzzy membership functions is directly tied to the goal of the control system. In a liquid level controller, the goal is to drive the liquid level to and maintain the level at a given setpoint in a minimum amount of time; that is to drive E and ΔE to zero in the shortest time possible. Thus, the fitness function must in some way include a concept of time.

Early in the research project, in discussing potential fitness functions for membership function selection in the liquid level problem, we considered numerous methods. For example, we discussed the ideas of minimizing energy consumption, of ensuring a critically damped system, and of never allowing the system to move further from the desired state. However, these discussions always seemed to return to plots of the liquid level versus time like that shown in Figure 6.5. In this figure, the vertical lines depict an error, E, at any particular time. The objective is not to have a controller that minimizes any individual error term, rather it is to minimize the sum of these error terms. Thus, the choice of fitness function for the liquid level problem must capture this idea:

$$f=\sum_{j=0}^{t_{max}}(E_j)^2$$

where f is the fitness function value, E is the error (difference between the current value of h and the setpoint defined for h), and t_{max} is the maximum time for which the computer simulation of the liquid level system is to be run. This definition of fitness function allows the GA to discover membership functions for effective control of the liquid level system from a given initial condition. However, the goal was to control the liquid level system beginning from **any** initial condition.

Thus, the fitness function described in Equation 6.2 above was extended to capture the idea that any initial condition could be considered. The second iteration of the fitness function used for the liquid level controller was:

$$f=\sum_{i=1}^{M}\sum_{j=0}^{t_{max}}(E_{i,j})^2$$

where M is the number of initial condition cases considered. Unfortunately, since the liquid level system can be in any of an infinite number of initial states, every initial condition cannot be considered. However, when the initial condition cases are selected carefully to represent the extreme regions of state space (maximum and minimum values of E and ΔE), an efficient and robust controller can be developed.

The only apparent problem associated with the fitness function of Equation (6.3) is that it is a value that we seek to minimize. As you recall, the GA described in Chapter 5 seeks to maximize fitness. Thus, the fitness function actually used in this effort (and used in the design of many controllers introduced later in this book) is:

$$f=K-\sum_{i=1}^{M}\sum_{j=0}^{t_{max}}(E_{i,j})^2$$

where K is a large constant selected to ensure that the value of f remains positive. By employing the fitness function of Equation 6.4 above, a GA was able to design a liquid level controller that was more efficient than the controller presented in Chapter 2.

6.3 RESULTS

A GA was used to select the membership functions for the liquid level fuzzy controller described in Chapter 2. When the coding scheme described in the previous section is employed, the GA must search a space that contains 2^k possible solutions where k is the length of the bit strings. Thus, in this problem the search space contained $2^{88}=3.09*10^{26}$ possible solutions to the problem; no wonder selecting high-quality membership functions is such a time-consuming task. After having sampled an amazingly small portion of the search space (approximately 3200 of the $2^{88}=3.09*10^{26}$ possible points), the GA was able to select membership functions that provided for better control than those defined by the authors.

Figures 6.5 and 6.6 represent comparisons of the genetic algorithm-designed fuzzy logic controller (GA-FLC) to the liquid level FLC developed by the authors (AD-FLC) for two sets of initial conditions. Figure 6.5 represents an initial condition that appeared explicitly in the fitness function employed by a GA to tune the controller. Thus, this figure depicts the performance of the GA-FLC on a set of initial conditions for which it was specifically trained. Naturally, the GA-FLC performs well in this instance. Perhaps a better indication of the GA-FLC's robustness is its ability to perform in conditions for which it has not been specifically trained. Figure 6.6 represents an initial condition that did not appear specifically in the fitness function employed by a GA. As in the previous case, the GA-FLC outperforms the AD-FLC, thereby indicating that the GA-FLC is robust. And, in fact, the GA-FLC outperforms the AD-FLC across a spectrum of initial condition cases.

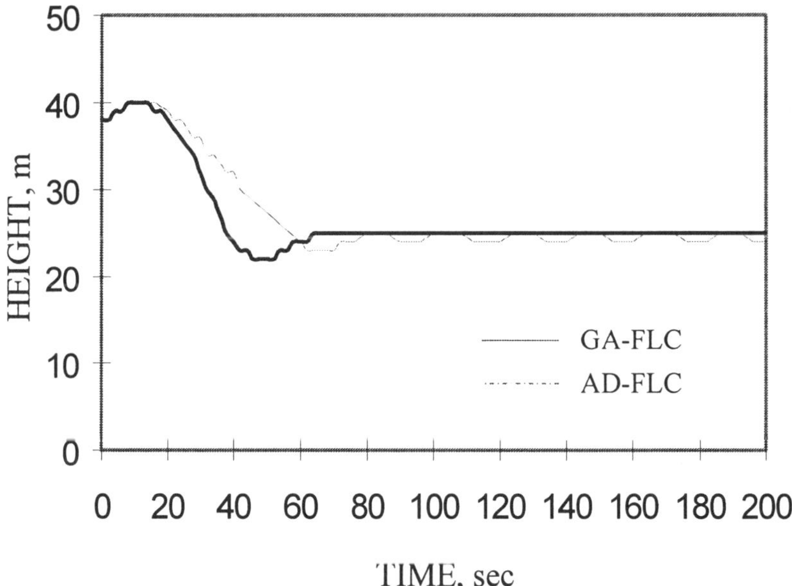

Figure 6.5 - The GA-FLC outperforms the AD-FLC for an initial condition that appeared in the fitness function.

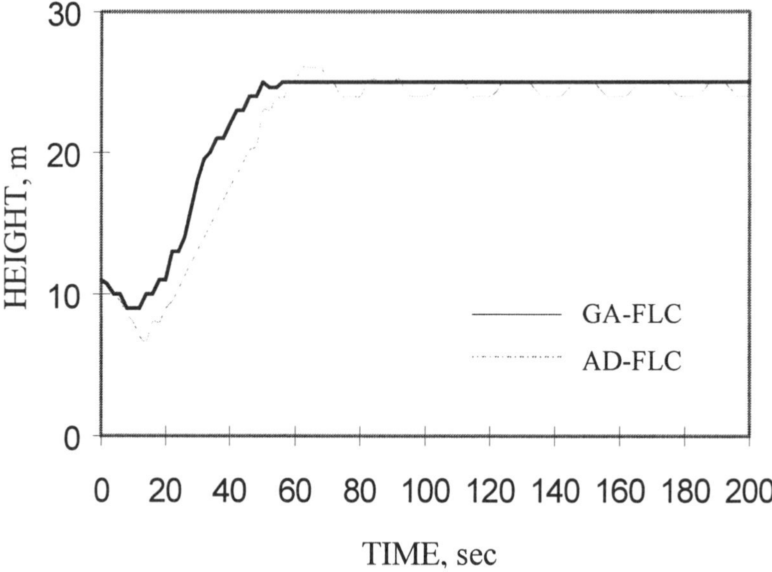

Figure 6.6 - The GA-FLC outperforms the AD-FLC for an initial condition that did not appear in the fitness function.

6.4 REVIEW/PREVIEW

In this chapter we have described the details of using a GA to define membership functions that improve the efficiency of a liquid level fuzzy control system. The two main issues discussed are the coding scheme and fitness function definition. The coding scheme introduced in Chapter 5 for a curve-fitting example was applied with little to no alteration. The fitness function was defined based on the objectives initially set forth for the controller in Chapter 2. With an appropriate coding scheme and a reasonable fitness function the GA was able to select membership functions that produced a controller that was more effective than the control system originally developed by the authors.

When we first implemented the approach presented in this chapter to membership function selection, we realized that this technique produced better controllers in substantially less time than the trial-and-error process previously employed. In fact, this approach effectively eliminates the need for a developer to use trial and error; it allows a GA to efficiently search the space of possible fuzzy systems, and as we saw in Chapter 5 GAs are capable of rapidly locating quality points in the search space. We immediately began to consider whether the approach was applicable to other control systems; to the selection of rule sets; to the selection of portions of rules sets. The answer to all of these questions, we learned, is a definite yes.

Chapters 7 and 8 demonstrate the extension of GA methods for the design of fuzzy controllers to other systems. The remainder of this book demonstrates some of the other ways GAs can empower fuzzy systems. As we were to discover, our quest for efficient and powerful fuzzy control systems was to be an amazing journey.

Chapter 7

GA-Tuned Cart-Pole Control

7.1 INTRODUCTION

At this juncture we have described three fuzzy systems: one for liquid level control, one for manipulating a cart-pole system, and one for a satellite rendezvous problem. Furthermore, we have introduced genetic algorithms and used them to define the membership functions for our liquid level controller. As we discussed in Chapter 6, there are two aspects to the design of fuzzy systems. There is the selection of membership functions, and there is the selection of high-quality rules. For the liquid level controller, selecting the rules was not an issue because they were intuitively apparent, and thus the focus of Chapter 6 was the selection of membership functions. However, high-quality rules are not always readily apparent for all problems.

As was discussed in Chapter 3, where we initially developed a controller for the cart-pole system, some of the rules were difficult to define because the system is coupled. There are conditions for which the current cart position suggests one force while the current pole angle suggests another. Previously, these apparent conflicts were resolved via a trial-and-error process in which the developers experimented with rules until a rule set was found that caused the system to perform acceptably. With the introduction of GAs, this trial-and-error process is no longer necessary. The current chapter describes the use of a GA to select the rules to be used in the cart-pole fuzzy controller.

Recall that in the cart-pole system, a cart is free to translate along a one-dimensional track while a pole balanced on the cart is only free to rotate in the vertical plane of the cart and track. A force, F, is applied to the center of mass of the cart. The objective of the control problem is to apply appropriate forces to the cart until it is motionless at the center of the track with the pole in a vertical position; this goal is to be accomplished in as short a time as possible. A schematic of the cart-pole system is shown in Figure 7.1. The controller must be able to achieve the control objective from any initial condition.

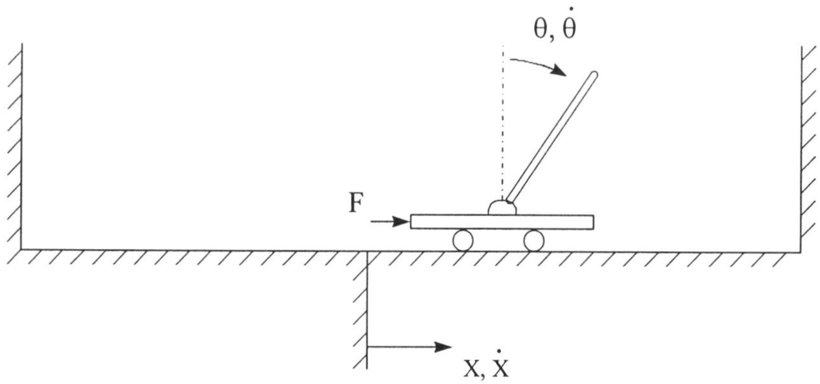

Figure 7.1 - The cart-pole system is a simple model of an inherently unstable multiple-output, dynamic system. Such systems are present in many balancing situations, e.g., two-legged walking and stabilizing a rocket during vertical takeoff.

A fuzzy cart-pole controller was designed and demonstrated in Chapter 3. This fuzzy controller used four condition variables: (1) the position of the cart, x, (2) the velocity of the cart, $\dot{x}$, (3) the angle of the pole, θ, and (4) the angular velocity of the pole, $\dot{\theta}$. The controller used the values of these four conditions to compute an appropriate action F from one of seven discrete linguistic possibilities: (1) NEGATIVE BIG, (2) NEGATIVE MEDIUM, (3) NEGATIVE SMALL, (4) ZERO, (5) POSITIVE SMALL, (6) POSITIVE MEDIUM, and (7) POSITIVE BIG. These seven choices were, of course, represented by fuzzy membership functions. The controller included membership functions used to describe each of the condition and action variables, and rules that prescribe an action for any condition existing in the cart-pole system (as described by a chosen number of linguistic terms).

In the cart-pole problem, the appropriate action for many of the rules is readily apparent, and thus required little effort to define. However, there were approximately 36 rules that required a trial-and-error procedure to define because they were characterized by conditions on the cart and pole that suggested conflicting actions. This trial-and-error process took a long time to accomplish (approximately four man-months; thank goodness for students). However, the process did result in a high-quality, effective rule set, and thus an efficient controller. There are two issues to be considered here: (1) the performance capabilities of the fuzzy system and (2) the time it takes to develop the controller. This chapter addresses the second of these issues because the fuzzy membership functions and the rules used previously in Chapter 3 have already been optimized using a trial-and-error process.

In the current chapter, two distinct tasks are accomplished. First, a GA is used to re-develop a rule set for the cart-pole system wherein the action for each rule can be one of any of the seven linguistic terms used in Chapter 3. In this effort, the goal is to produce a fuzzy controller that is as effective as the optimized controller from Chapter 3. The benefit of using a GA in this first task is that the rule set can be developed in far less time than it took us initially. Second, a GA is provided with an expanded set of linguistic terms from which to choose the rule values. This is done to assist in the presentation of a new type of coding, tertiary coding. As a consequence of the new coding, a slight alteration to the basic GA of Chapter 5 is described. The result is a controller that is more effective than the cart-pole controller of Chapter 3.

7.2 RULE SELECTION IN LESS TIME

The cart-pole problem of Chapter 3 is a challenging control problem and a benchmark for testing process control strategies. Since a near-optimum fuzzy controller has been developed for the cart-pole system, this problem provides an effective test-bed for investigating the ability of a GA to re-discover the appropriate action to be associated with each rule. Fuzzy membership functions have already been defined for the condition and action variables and these membership functions are used here. Since the condition variables and the associated fuzzy membership functions are defined, and since the approach we have adopted is to write a rule for every possible combination of condition variables, the number of rules is determined. In this case, there are 108 rules to be written because x is described using four linguistic terms, $\dot{x}$ is described using three linguistic terms, θ is described using three linguistic terms, and $\dot{\theta}$ is described using three linguistic terms; (4)(3)(3)(3)=108. There still remains the task of selecting the appropriate action to be associated with each of the 108 rules.

All of the rules in the cart-pole system are not readily apparent. For example, consider the situation for which the following conditions exist: (1) x is POSITIVE SMALL (just to the right of center), (2) $\dot{x}$ is NEAR ZERO (the cart velocity is small), (3) θ is POSITIVE (the pole is to the right of vertical), and (4) $\dot{\theta}$ is NEAR ZERO (the angular velocity of the pole is small). This situation is depicted in Figure 7.1. Even though the membership functions for the seven potential actions are defined (as shown in Figure 7.2), the appropriate action is not readily apparent. The conditions on the cart (x and $\dot{x}$) seem to suggest a force of NEGATIVE SMALL, while the conditions on the pole (θ and $\dot{\theta}$) seem to suggest a force of

POSITIVE SMALL. Naturally, the developer can attempt to resolve the conflict by reasoning about the system (in this case, balancing the pole is a priority so a POSITIVE FORCE is applied). Such reasoning often fails. In our experience, the decision almost always becomes a trial-and-error process. Moreover, we have worked with systems that we did not understand as well as we do the cart-pole system (the helicopter system presented in Chapter 18 is an example). In these systems, the necessary analysis is not feasible. Thus, the determination of effective actions for each of the rules is fertile ground for applying a GA.

When using GAs for rule selection, there are still the two issues of coding and fitness function determination that must be addressed. Fortunately, the techniques presented in Chapters 5 and 6 are again applicable. The details of their implementation follow.

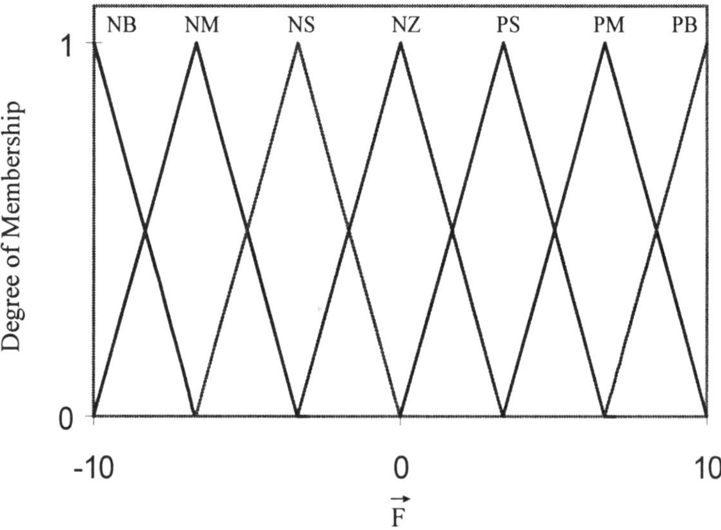

Figure 7.2 - The membership functions for F set forth in Chapter 3 are used here. These membership functions provide the developer with seven linguistic terms from which to choose an appropriate action.

7.2.1 Coding

When no rules are pre-selected because they are obvious, there are 108 conditions for which one of seven linguistic terms is to be selected. Thus, there exist $108^7 = 1.713*10^{14}$ possibilities for the rule set. The problem of coding involves developing a mechanism in which an entire rule set can be encoded as a string of characters. In this case, the characters will be bits.

We will employ a form of the concatenated, mapped, unsigned binary coding described in previous chapters. In this coding, substrings represent individual parameters (in this case individual rule actions), which are linearly mapped, and concatenated to form strings that represent complete solutions (entire rule sets). One of seven terms is to be selected for each of 108 conditions. Thus, each substring must represent one of the seven possible action values. The following encoding is used:

string	interpretation
000	NEGATIVE BIG
001	NEGATIVE MEDIUM
010	NEGATIVE SMALL
011	NEAR ZERO
100	NEAR ZERO
101	POSITIVE SMALL
110	POSITIVE MEDIUM
111	POSITIVE BIG

Note that in this encoding, there are two strings representing NEAR ZERO. This is because the seven choices for force require at least three binary digits (two binary digits allow for the representation of only four choices). In general, this duplication is not desirable because the rule values are biased to NEAR ZERO. However, in this instance, the bias of the solutions associated with the duplication is outweighed by the sheer simplicity of the coding representation. In Chapter 9 we will demonstrate an alternative solution in a more difficult instance of the cart-pole system.

Given the above representation of the individual rule actions, a complete rule string is easily represented using a bit-string composed of 108 individual three-bit substrings. Thus, the bit-strings used in this coding are of length 108*3 = 324. As a point of interest, this is a rather long bit string representing $2^{108} = 3.24*10^{32}$ possible solutions. This calculation reveals another problem associated with the duplication in the coding scheme. The GA must search a space with more points than is actually required in the problem, $3.24*10^{32}$ versus the $108^7 = 1.713*10^{14}$ distinct possibilities.

7.2.2 Fitness Function

Once an encoding scheme has been declared, the remaining task is to define a fitness function that will drive the GA to high-quality solutions. In Chapter 6 we presented a fitness function used to develop high-quality membership functions in a liquid level controller. In that instance, the objective function was meant to provide high fitness values to controllers that drove the liquid level to a predefined setpoint in the least amount of time possible. The goal of the cart-pole controller is to drive the cart to the center of the track (while keeping the pole balanced) in the shortest time possible. Thus, the fitness function for the selection of rules in the cart-pole system is:

$$f = \sum_{j=0}^{j=M} \sum_{i=0}^{t_{max}} w_1 (x - x_{desired})^2 + w_2 (\theta - \theta_{desired})^2 \quad (7.1)$$

where t_{max} is the amount of time for which all cart-pole simulations are to be run and M is a pre-defined number of initial condition cases (we used four initial condition cases selected from extremes of the state space, i.e., x = NEGATIVE BIG, $\dot{x}$ = NEGATIVE, θ = NEGATIVE, and $\dot{\theta}$ = NEGATIVE), $x_{desired}$ and $\theta_{desired}$ are the setpoints for x and θ, respectively, and w_1 and w_2 are weighting values. The weighting values are necessary because the values of x are generally an order of magnitude larger than the values of θ. For the pole angle to be considered as important as the cart position, w_2 must be an order of magnitude larger than w_1.

An important point about the fitness function above is that fit individuals have low scores. In our previous discussions of genetic algorithms, we declared fitness to be a value that was to be maximized because of the reproduction operator being used. There are reproduction operators that can be used in situations such as the one above in which the fitness function actually represents a form of error that is to be minimized (Goldberg, 1989). One such operator is the tournament selection operator. The operator is implemented as follows. A subset of the strings in the current generation, say p strings, is selected for a tournament competition. The fitness values of these p strings are compared, the top performer is placed in the mating pool (which serves as a temporary file) where they await the operation of crossover. In a maximization problem, the top performer is the string with the largest fitness function value; in a minimization problem, the top performer is the string with the smallest fitness function value. These tournaments are continued until the mating pool is full, i.e., until popsize tournaments have been conducted. Thus, with the use of the tournament selection operator, the fitness function can just as easily be a value that is to be minimized.

7.2.3 Results

The coding scheme and fitness function defined above was used to guide a GA search for the actions associated with a rule set employed by a fuzzy controller for the cart-pole system. The GA efficiently considered the possible solutions and rapidly converged to a set of actions for a rule set equivalent to the one presented in Chapter 3. Recall that the rule set presented in Chapter 3 required approximately four man-months to develop. The GA's search was conducted overnight on a 486-based personal computer. Naturally, there was some overhead associated with the computer programming required to implement the GA's search, but it was on the order of a single day. Thus, a GA can be used to determine effective rule actions for fuzzy systems. The time necessary for determining effective rules can be reduced dramatically without a decrease in the effectiveness of the fuzzy system.

In the next section, the problem of rule selection for the cart-pole system is made more difficult through the inclusion of additional linguistic terms. As a consequence of increasing the number of linguistic terms available for describing the force, a new and more efficient coding scheme is used. The GA proves to be equally effective in the more complex system.

7.3 AN EXPANDED RULE SET AND A NEW APPROACH TO CODING

The precision of a controller can be limited when we describe the possible actions with a finite number of linguistic terms. This is why we introduced the real-valued consequent approach in Chapter 4. However, since so many fuzzy systems have adopted the "traditional approach," we will consider it further here. In the previous section, we addressed the cart-pole problem as described in Chapter 3 where the action associated with each rule was to be one of seven choices. One approach to achieving more precise control of a system is to increase the number of linguistic terms from which the actions can be chosen. However, when the number of action choices is increased, selecting the rule set becomes more difficult simply because there are more distinctions to be made.

Previously (Chapter 3 and section 7.2) we defined membership functions representing seven linguistic values of force applied to the cart. Here, we examine the ramifications of increasing this number from seven to nine. Consider the membership functions for force shown in Figure 7.3 in which the following nine linguistic terms are defined: (1) NEGATIVE VERY BIG, (2) NEGATIVE BIG, (3) NEGATIVE MEDIUM, (4) NEGATIVE SMALL, (5) NEAR ZERO, (6) POSITIVE SMALL, (7) POSITIVE MEDIUM, (8) POSITIVE BIG, and (9) POSITIVE VERY BIG. Notice in the figure that the limits of force have not been extended, so that the nine membership functions shown cover the same range of force values as the seven membership functions shown in Figure 7.2. Thus, the meaning of the linguistic terms have changed somewhat.

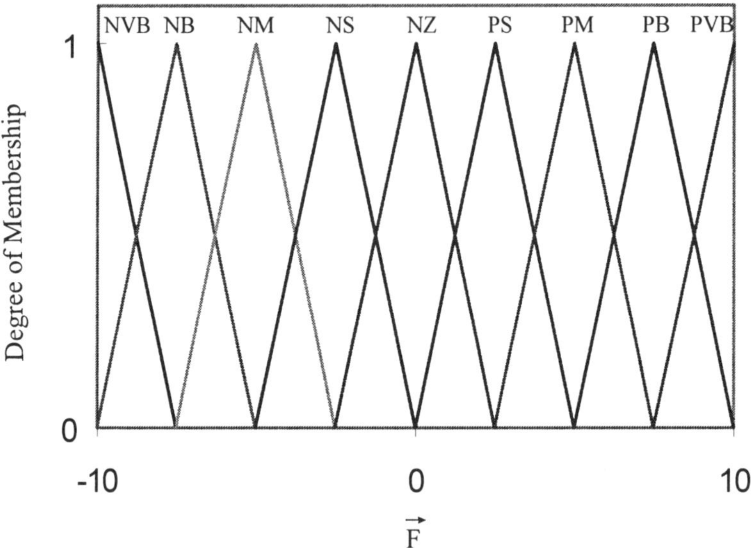

Figure 7.3 - With the availability of the nine linguistic terms for describing a value of force, more effective controllers are possible. However, the problem of rule selection becomes more difficult since there are now 108^9 or approximately $2*10^{18}$ possible rule actions to be considered.

7.3.1 Coding

At first glance, the addition of two new choices for the actions has not changed anything that should make us reconsider our use of the concatenated, mapped, unsigned binary coding described above. However, upon further consideration, perhaps we should make some adjustments. Consider the use of binary strings of length two to represent the nine choices we have for each rule. If we use binary strings of length two, we can without duplication directly represent four choices; with binary strings of length three we can represent eight choices; with binary strings of length four we can represent sixteen choices. Thus, binary coding requires binary strings of length four be used to represent the nine choices for the linguistic variables describing force to be associated with each rule. We will have seven "unused" bits that will be filled with duplicate selections. Recall we faced this situation in the previous section in which we duplicated one linguistic descriptor of force. However, all of the problems associated with duplication of solutions are now magnified. Consider for instance the size of the search space. With the declaration of nine choices for the rules, there exist approximately $2*10^{18}$ possible rule sets. However, with the binary coding being discussed, the GA would be faced with a search space consisting of $2^{108*4} = 1.109*10^{130}$ possible solutions (certainly some of them are duplicates, but they are still possibilities). Thus, it is prudent to consider an alternative coding scheme.

With nine rule actions to choose from, a tertiary coding is natural. Consider a coding scheme in which base three representations are used instead of binary or base two representations. If such an alphabet is used, then the following table lists the mapping from base three space to rule value space:

string	interpretation
00	NEGATIVE VERY BIG
01	NEGATIVE BIG
02	NEGATIVE MEDIUM
10	NEGATIVE SMALL
11	NEAR ZERO
12	POSITIVE SMALL
20	POSITIVE MEDIUM
21	POSITIVE BIG
22	POSITIVE VERY BIG

Notice in the above representation no duplication of solutions is required; each string of characters maps directly and uniquely to a single choice of action. Thus, the problems associated with duplication in the coding scheme discussed in the previous section are no longer relevant. Additionally, a GA is now faced with a search space of size $3^{108*2} = 1.143*10^{103}$ which is far smaller than that created using a binary representation. Aside from the leverage gained by reducing the size of the search space, it is generally better to avoid duplication when possible because of the bias introduced; strings that are represented with more than one substring represent a larger percentage of the possible solutions. Thus, they are more likely to exist in the population, whether they are quality solutions or not. This bias can cause the GA to converge to sub-optimal answers or to take longer to converge to the best answers.

Given the expansion of our coding scheme from binary representation to a ternary alphabet, we must ensure that the operators employed by the GA can still be used. Reproduction biases successive populations of strings toward highly fit strings. It employs the fitness function values to drive its choice of strings to copy in the mating pool with our tournament selection. Reproduction is not affected by the extension to a base three representation. Crossover still involves the combining of strings. It selects two strings from the mating pool, cuts the strings at a randomly selected site, and exchanges the tails of the strings to form two new strings. Like reproduction, crossover is not affected by the use of a ternary alphabet. Mutation, however, is a different story. Mutation involves the random alteration of string values. In our binary strings, it meant changing a 0 to 1, and a 1 to a 0. In our current coding, we must accommodate not only the familiar 1 and 0 characters, but also the newly introduced two character. Thus, a simple expansion of mutation (and others are presented in the reference by Goldberg, 1989) is to change a 0 to a 1, change a 1 to a 2, and change a 2 to a 0. With this modification of the mutation operator, the GA is fully capable of working with the ternary coding scheme which results in 108, two-character substrings concatenated to form strings of length 216.

7.3.2 Fitness Function

The changes in the coding scheme have in no way altered the desired goal of the controller. Therefore, it is reasonable to assume that the fitness function defined in the previous section is again applicable. The fitness function of Equation (7.1) is employed here.

7.3.3 Results

In the previous section we demonstrated the ability to locate an already defined optimum rule-set in less time than needed by a human developer (at least in less time than it took us). The results of this section should provide a little something different. First of all, we do not know what the optimum controller for the cart-pole system is when we have nine values available from which to choose an action. Thus, the performance of the nine-action term GA-FC is directly compared to the performance of the seven-action term GA-FC developed in the previous section. As can be seen from Figure 7.4, the nine-action GA-FC is more robust. As a note, the initial conditions depicted in Figure 7.4 were not one of the four initial condition cases appearing in the fitness function. Thus, this figure represents the performance of the GA-FC on an initial condition case for which it was not specifically trained.

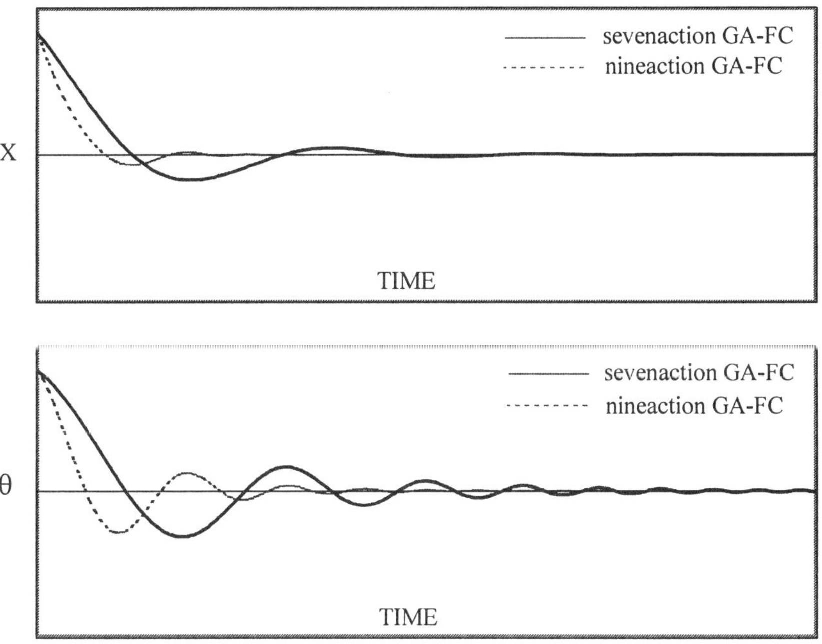

Figure 7.4 - The optimized nine-action GA-FC is more efficient than the optimized seven-action GA-FC. Thus, it is apparent that expanding the values from which an action can be chosen gives a more effective controller. However, without an effective search technique such as a GA, determining optimum rule values is a daunting task.

7.4 INTRODUCTION OF ERROR

Up to this point in the book we have developed fuzzy systems for controlling environments simulated on a computer. This is nice for us because we have been working with the proverbial "perfect world." As shown in Figure 7.5, the computer simulations have been used to provide feedback information to the controller concerning the state of the problem environment. In this situation, the simulation is used both to model the problem of interest (the liquid level, cart-pole, and rendezvous systems) and to provide sensory information to the controller. Thus, the sensory information has been free of error (except roundoff error). The sensory information provided in the simulated environments we have worked with to date is not representative of the sensory information available in real-world systems for a number of reasons. First, the models are considered to be perfect when real-world models are not always perfect. Second, there are no mistakes in the simulated sensory information; no outliers. Third, the only error is roundoff error which is very small.

Even though these simulated systems have provided us an effective vehicle for presenting our ideas concerning fuzzy systems and GAs, we need to address the issue of error in sensory information if we are to be prepared for the real-world systems addressed later in this book. Thus, section 7.4 is meant to examine the effectiveness of GA-FCs in the presence of error.

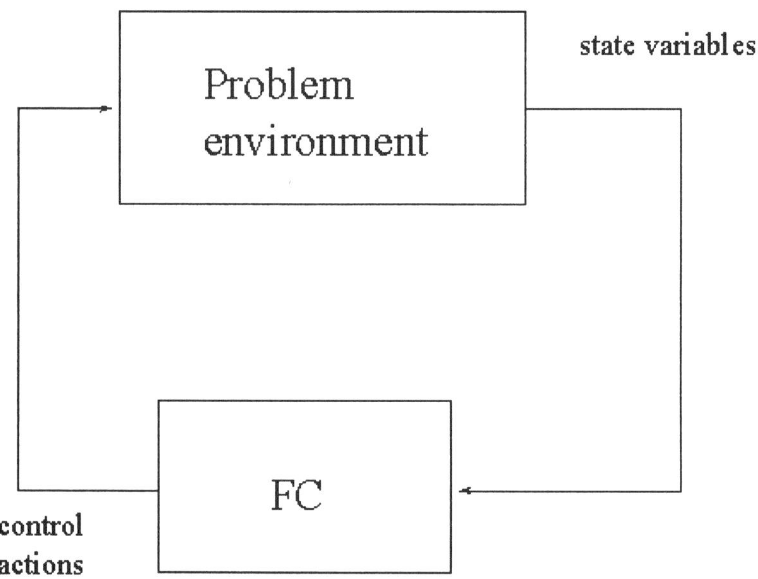

Figure 7.5 - The simulated environments addressed to this point take advantage of the availability of perfect sensory information when describing the status of the state variables to the controller.

The presence of sensor error is easily simulated in the cart-pole system we are considering. The values of x, $\dot{x}$, θ, and $\dot{\theta}$ are simply altered with the addition of various amounts of Gaussian noise. Then, a GA is used to select optimum rules for systems with various amounts of sensory error. Figure 7.6 shows the performance of a GA-FC for the cart-pole system in the presence of 5 and 10 percent noise. Notice that the performance of the optimized controller degrades as the level of noise is increased. This is exactly the effect one would expect with the introduction of noise. The point here is that fuzzy controllers can tolerate substantial amounts of noise.

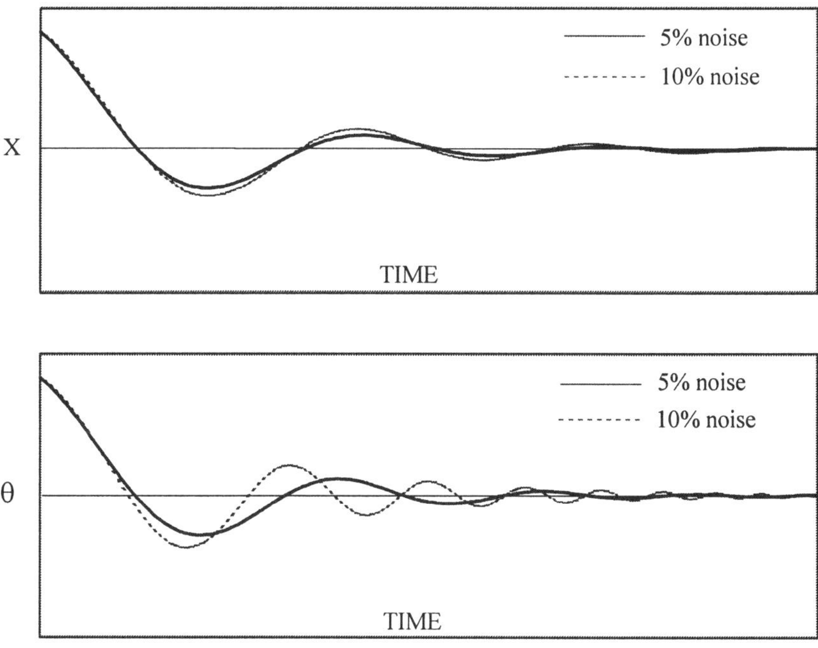

Figure 7.6 - The performance of a fuzzy controller degrades as the magnitude of the sensory error is increased. Fuzzy controllers cannot completely overcome the presence of noise even when used in conjunction with a GA.

7.5 REVIEW/PREVIEW

This chapter has demonstrated the effectiveness of using GAs for selecting optimum rule sets employed by a fuzzy cart-pole controller. Two specific cases were considered. In the first, a GA was used to select the optimum values for a seven-action fuzzy controller. In this instance, the GA located the optimum rule set determined by the authors in a four-month trial-and-error process. The GA was able to find the rule set overnight. Second, a GA was used to determine an expanded rule set in which the developer was free to choose one of nine actions for each of 108 rules in the cart-pole controller (as opposed to the seven choices available in the controller presented in Chapter 3). In this instance a GA defined a rule set that allowed more-effective control of the cart-pole system. Finally in this chapter, the concept of error in sensory information was introduced and accounted for in a simulated environment.

The next chapter addresses the use of a GA to improve the performance of our satellite rendezvous controller introduced in Chapter 4. We will address the issue of using a GA to select real-valued consequent values, and we will also see that a GA can be used to simultaneously locate rules and membership functions.

Additionally, we will discuss the ramifications of using GAs to optimize controllers to fuzzy implication operators.

REFERENCES

Goldberg, D. E. (1989). *Genetic algorithms in search, optimization, and machine learning*. Reading, MA: Addison-Wesley.

Chapter 8

Satellite Rendezvous Using a Genetic Algorithm

8.1 INTRODUCTION

Genetic algorithms are useful for tuning fuzzy systems for several reasons. We have demonstrated some of the reasons in previous chapters. For instance, one inviting aspect of the GA is that it does not require derivative information that would be difficult and computationally expensive to obtain in many control situations. Second, a GA is flexible enough to be used both for tuning membership functions and for selecting rules. We will demonstrate additional reasons in the current chapter. We will initially demonstrate the effectiveness of using a GA to select the real-valued consequents of rules in the satellite rendezvous system. In fact, we will show that a GA can be used to simultaneously select real-valued consequent rules and adjust membership functions. Also, we will discuss the effect that precise optimization of fuzzy systems has on the selection of implication operators.

In the previous two chapters we used a GA to improve the performance of both the liquid level controller of Chapter 2 and the cart-pole controller of Chapter 3. In this chapter we will turn our attention to the satellite rendezvous system presented in Chapter 4. Recall that in the satellite rendezvous system we implemented a real-valued consequent approach. Thus, we will use a GA to select rules of this form. Additionally, we will show that the tasks of adjusting membership functions and selecting rules do not have to be accomplished independently of one another. In fact, they are best tackled together (Karr, Fleming, and Vann, 1994).

There is much discussion in the fuzzy community as to which implication operator is best. We have introduced two implication operators, the min-max operator used in the liquid level controller and the multiplication operator used in the rendezvous system. In this chapter we will demonstrate that for the satellite rendezvous system (and probably for most systems), when the fuzzy membership functions and the applicable rule set are selected simultaneously, the choice of implication of operator is not critical. Of course, the controller must be tuned with a particular implication operator.

8.2 REVIEW OF THE RENDEZVOUS SYSTEM

In Chapter 4 we introduced a satellite rendezvous system. Here, we will briefly review the details of the problem and of the controller developed earlier. The rendezvous system consists of two spacecraft, one actively performing the rendezvous, the other passively orbiting the earth. A relative coordinate system is used to describe the state of the rendezvous system since the sensors on board provide relative information. The equations of motion governing the system are:

$$\ddot{x} - 2\omega\dot{y} = \frac{F_x}{m} \tag{8.1}$$

$$\ddot{x} - 2\omega\dot{y} = \frac{F_x}{m} \tag{8.2}$$

$$\ddot{z} + \omega^2 z = \frac{F_z}{m} \tag{8.3}$$

where ω is the angular velocity of the relative coordinate system origin rotating about the earth; F_x, F_y, and F_z are thrust forces; m is the mass of the chaser vehicle; and x, y, and z are the relative coordinate positions. The goal of the associated fuzzy controller was to complete the rendezvous (drive the values of x, y, and z) to zero in the shortest time possible.

The fuzzy system developed to control the rendezvous system consisted of six condition variables: (1) the x position coordinate, x, (2) the y position coordinate, y, (3) the z position coordinate, z, (4) the x-velocity, $\dot{x}$, (5) the y-velocity, $\dot{y}$, and (6) the z velocity, $\dot{z}$. These six condition variables were each described with six linguistic terms: (1) NEGATIVE-BIG (**NB**), (2) NEGATIVE-SMALL (**NS**), (3) NEGATIVE-CLOSE (**NC**), (4) POSITIVE-CLOSE (**PC**), (5) POSITIVE-SMALL (**PS**), and (6) POSITIVE-BIG (**PB**). The membership functions shown in Figure 8.1 were used to describe each of the six condition variables; there was no reason to differentiate between the x, y, and z directions. The fuzzy controller used information describing the state of these six condition variables to prescribe an action for each of three action variables: (1) thrust in the x-direction, F_x, (2) thrust in the y-direction, F_y, and (3) thrust in the z-direction, F_z. These three action variables were not described with linguistic terms, rather they were described using a real-valued consequent. Therefore there were no membership functions for the action variables.

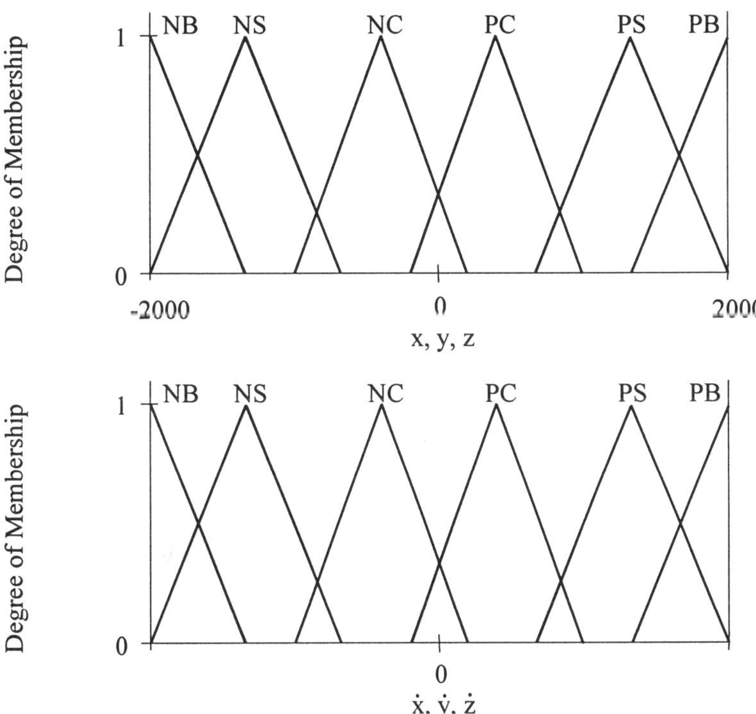

Figure 8.1 - The twelve membership functions shown above, six describing x, y, and z, and six describing $\dot{x}$, $\dot{y}$, and $\dot{z}$, were adequate to allow a fuzzy controller to effectively manipulate the rendezvous system.

Aside from the basic details described above, it is important to remember that the coupling effects apparent in the rendezvous problem were considered with a supplementary rule set. The rendezvous controller employed 36 rules each describing appropriate actions for F_x, F_y, and F_z. The coupling in the system was then addressed with the resulting 3 * 36 = 108 rules and with 16 supplementary or coupling rules. The result was a rather robust set of 124 rules that were used in conjunction with the fuzzy membership functions shown in Figure 8.1 to manipulate the rendezvous system.

8.3 GENETIC ALGORITHM IMPROVES THE SYSTEM

The rendezvous controller of Chapter 4 was developed without the aid of a GA to locate membership functions and rules. Since a real-valued consequent approach was used there are an infinite number of rule sets from which to choose. Thus, there is good reason to believe that the controller presented in Chapter 4 can be improved. In this section, we will use a GA to locate rules and membership functions that improve the performance of the rendezvous controller. Again, the two issues to be addressed are the coding scheme and the fitness function.

8.3.1 Coding Scheme

The concatenated, mapped, unsigned binary coding employed in previous chapters is quite robust and can represent both rules and membership functions. It will also be used here to represent these parameters taken together.

With the linguistic terms selected to describe the condition variables in the rendezvous system (triangular membership functions were used as shown in Figure 4.2), there are ten triangle base points that must be defined for each condition variable, and thus there are 60 points that must be defined to completely describe the membership functions. With four bits each allotted to each base point, the substrings representing an entire set of membership functions is 240 bits long.

The real-valued consequents associated with the rules do not require any special attention; they are continuous variables existing on a fixed range. Thus, these values can also be represented using our familiar concatenated, mapped, unsigned binary coding. The substrings representing the individual rules are simply concatenated to the substring representing the membership functions. Recall that the rendezvous controller consists of 124 total rules (of the 124, 16 are coupled rules). Each rule was represented with a binary string of length three. The substring representing a complete rule set was therefore 372 bits in length. When combined, the substring representing a set of fuzzy membership functions and a substring representing a rule set were of length 612. This is an extremely long string length for the GA to work with. As an indication, consider that for this discretization, there are $2^{612} = 1.699 * 10^{184}$ possible solutions to the search problem.

8.3.2 Fitness Function

The goal of the rendezvous controller is to drive the values of x, y, and z to setpoint or desired values (to complete the rendezvous) in as little time as possible. This goal is quite similar to those defined for the liquid level and cart-pole systems. Thus, it is natural to consider the fitness functions set forth in the previous chapters. A reasonable fitness function for the rendezvous system would seem to be:

$$f = \sum_{j=0}^{j=M} \sum_{i=0}^{t_{max}} (x - x_{desired})^2 + (y - y_{desired})^2 + (z - z_{desired})^2 \tag{8.4}$$

However, upon closer inspection, such a fitness function is not adequate. The goal is to drive the relative distance between the chaser and target vehicles to zero. This distance, r, is defined by the equation

$$r = \sqrt{x^2 + y^2 + z^2} \tag{8.5}$$

There is a constraint on the problem: r must be driven to zero *gradually*. This distance should not be zero while there are large relative velocities; this means the chaser vehicle crashes into the target. Certainly, this is a concern especially if one recalls the critically damped behavior exhibited by the liquid level and cart-pole systems. A constraint must be added into the fitness function to dissuade the GA from crashing the vehicles. The constraint is a limitation to the magnitudes of the velocities ($\dot{x}$, $\dot{y}$, and $\dot{z}$) that can exist when the distance (r) is driven to zero. Goldberg (1989) suggests a penalty function for this type of constrained optimization. The penalty function used here is

$$f = [\sum_{j=0}^{j=M} \sum_{i=0}^{t_{\max}} (x - x_{desired})^2 + (y - y_{desired})^2 + (z - z_{desired})^2] + K \tag{8.6}$$

where K is a penalty that is proportional to the degree to which the constraint is violated. In this case,

$$K = 100{,}000 * (\dot{x} + \dot{y} + \dot{z}) \tag{8.7}$$

In this way, the magnitude of a fitness value is increased proportional to the magnitude of the relative velocities existing when the rendezvous is complete.

8.3.3 Results

A GA was used to select both the membership functions and the rule set employed by a rendezvous controller. The results are summarized in Figure 8.2. In this figure, the performance of the GA-FC is compared with the performance of the controller developed by the authors (AD-FC) and presented in Chapter 4. Notice that not only does the GA-FC complete the rendezvous in less time, but it does not allow the distance between the vehicles to go to zero when there are large velocities. In this case, the overdamping effect displayed by the GA-FC is necessary because of the constraints that restrict the system from allowing the distance r to go to zero if the chaser craft has a large velocity.

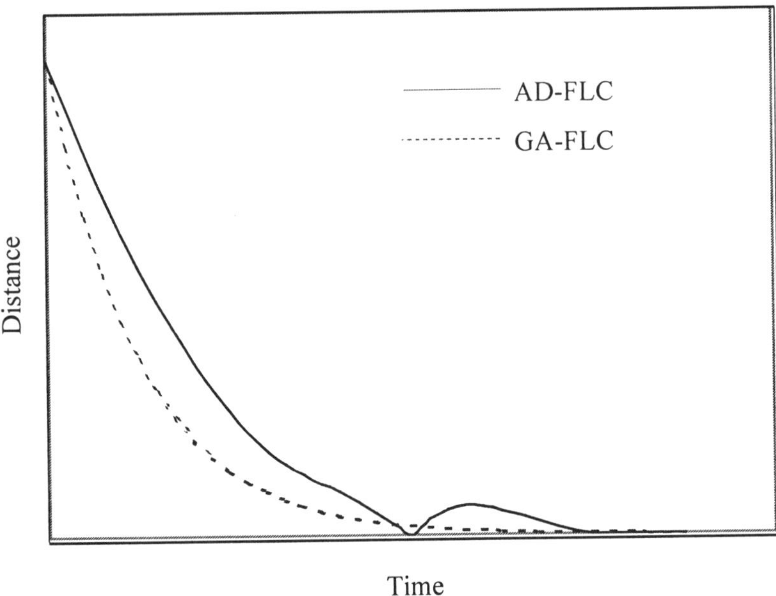

Figure 8.2 - The GA-FC completes the rendezvous in less time than the AD-FC presented in Chapter 4. In addition, the GA-FC prevents the “crash” that the AD-FC causes at t = ** seconds.

8.4 IMPLICATION OPERATORS

In our original presentation of the rendezvous controller in Chapter 4, we used a multiplication operator instead of the min-max implication operator we used in previous chapters. To gauge the effects of the two operators, we decided to compare the multiplication and min-max implication operators in the rendezvous problem. However, to ensure a fair comparison, we used a GA to design each controller. Figure 8.3 summarizes the results achieved. There is much discussion in the *fuzzy community* concerning the relative merit of implication operators. It is our experience that when a GA is used, the choice of implication operator is virtually inconsequential (Karr, Fleming, and Vann, 1994).

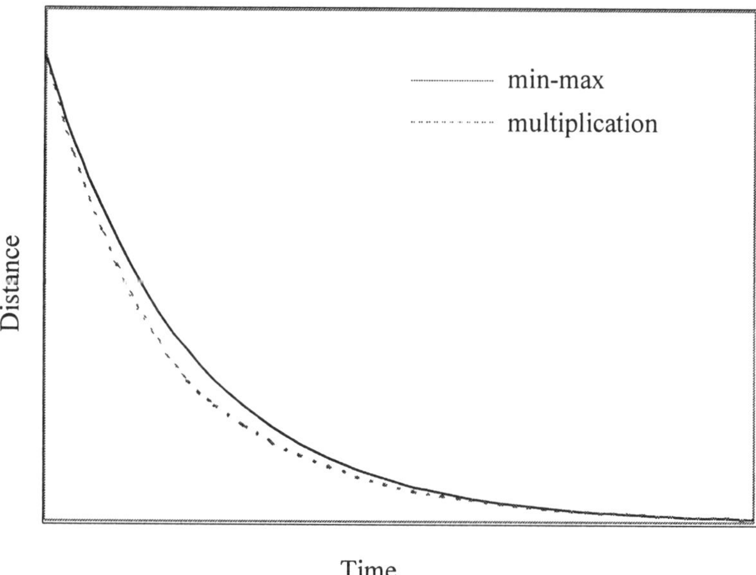

Figure 8.3 - The choice of implication operator appears to be inconsequential when a controller is optimized using a GA.

To gain a better understanding of this observation, we did a thorough investigation using numerous implication operators, membership function forms, and rule forms. A complete discussion of the results of this study appear in the report by Karr, Fleming, and Vann (1994). Briefly, the selection of implication operator did not seem to affect the performance of the controller when a GA was used to design the controller.

8.5 REVIEW/PREVIEW

In this chapter we have done several things. First, we demonstrated the ability of a GA to simultaneously select membership functions and rules, and to select real-valued consequent for the rules. Second, we introduced a fitness function applicable to constrained optimization problems. Finally, we discussed the effect of different implication operators on the performancc when GAs were used for optimization.

At this juncture we have used a GA to optimize both the membership functions and the rules for three different control systems: (1) a liquid level controller, (2) a cart-pole controller, and (3) a satellite rendezvous controller. It seemed that our development was complete. However, it was at this juncture in the research project that a whole new world opened up to us. In the next chapter we will open a door to this world by repeating the comment our boss made to us one morning: "that cart-pole controller works well, but what happens when the mass of the cart changes?" In the next chapter we will introduce the idea of and need for adaptive fuzzy controllers.

REFERENCES

Goldberg, D. E. (1989). *Genetic algorithms in search, optimization, and machine learning*. Reading, MA: Addison-Wesley.

Karr, C. L., Fleming, J. W., & Vann, P. A. (1994). *Aspects of genetic algorithm-designed fuzzy logic controllers* (Report of Investigations number 9515). Washington, DC: U.S. Department of the Interior, Bureau of Mines.

Chapter 9

Time-Varying Cart-Pole Controller

9.1 INTRODUCTION

To this point, we have presented a step-by-step approach for developing fuzzy systems, and we have introduced the genetic algorithm as a tool for assisting in the tuning of these fuzzy systems. In the current chapter we will introduce adaptive fuzzy control systems that have their rules and membership functions adjusted in "real-time" in response to changing problem environments. The adaptive fuzzy controllers introduced herein rely on a simulation (computer model) of the system being controlled. We use the cart-pole system of Chapters 3 and 7 to introduce these adaptive fuzzy systems.

In the cart-pole system there are parameters such as the mass of the cart, the length of the pole, and the coefficients of friction that remained constant. However, there is no compelling reason why these parameters cannot change, and when they do, this change can have a dramatic affect on the way the cart-pole system responds to forces applied to the cart. If sensory information about these system parameters is available, and these parameters are changing frequently, then the natural inclination is to incorporate them into the rule set. But, it should be noted that as we increase the number of parameters considered directly in the fuzzy system, the number of rules increases multiplicatively. For example, there are 108 rules in the cart-pole system as described in Chapters 3 and 7. If we were to expand the rule set so that it included five linguistic terms describing the mass of the cart, the rule set would require 5*108=540 rules. Rule sets that are unduly large are inappropriate for at least two reasons: (1) dramatic increases in the number of rules can slow the fuzzy system down and (2) when a rule set gets too large, human operators no longer feel comfortable with their understanding of the rules set. There is, however, at least one situation where the number of rules can be reduced: when parameters change slowly and are not directly measured. There are numerous occasions in which system parameters change infrequently where these changes are not directly measured with sensors and therefore cannot be included in the rules. Thus, it is worthwhile considering alternative approaches.

We can consider changes to the system parameters such as the mass of the cart, the length of the pole, and the coefficients of friction in the cart-pole system. The system parameters change less frequently than the state variables (x, $\dot{x}$, θ, and $\dot{\theta}$). Furthermore, it is assumed that no sensory information is available that measures the current values of the system parameters. Therefore, our original task of controlling the cart-pole system is greatly complicated. Now we are forced to do at least three things. First, we must continue to control the cart-pole system, termed the *problem environment*. We can accomplish this task using a fuzzy controller. Second, we must determine when and to what extent the system parameters have changed. When a system parameter such as the mass of the cart changes, the controlled system no longer responds in the same way as it did before. (If the mass of the cart is suddenly doubled, we must drive our cart differently). Third, once it has been determined that the problem environment has changed, we must update our fuzzy controller so that it can efficiently control the new problem environment.

9.2 TIME-VARYING CART-POLE SYSTEM

The concept of an adaptive controller is introduced using a modification of the cart-pole system described in Chapter 3 in which the cart's mass varies. After a brief review of the traditional system, the more complex, time-varying cart-pole system is introduced. Subsequent sections of this chapter address alternate approaches to controlling systems that have relatively slow changing parameters.

Recall that in the cart-pole system, the cart is free to translate along a one-dimensional track while a pole is free to rotate only in the vertical plane of the cart and track. A multivalued force, F, can be applied at discrete time intervals in either direction to the center of mass of the cart. A schematic of the cart-pole system is shown in Figure 9.1. The objective of the control problem is to apply forces to the cart until it is motionless at the center of the track with the pole balanced in a vertical position. This goal should be accomplished as quickly as possible. A block diagram of a very basic control loop is shown in Figure 9.2.

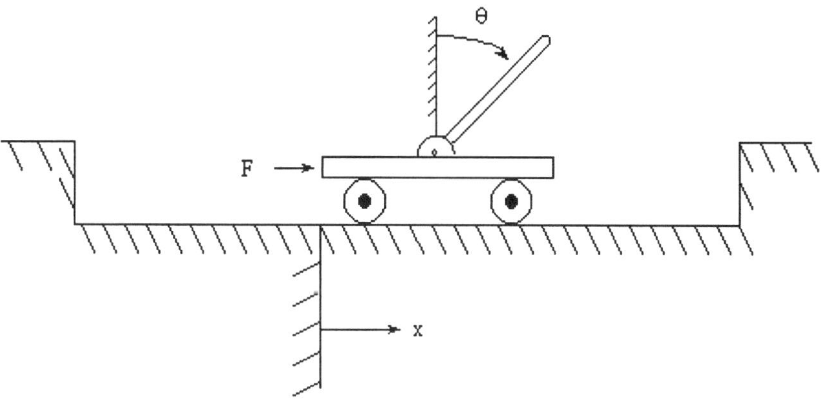

Figure 9.1 - The cart-pole system of Chapter 3 is used here to introduce adaptive fuzzy systems.

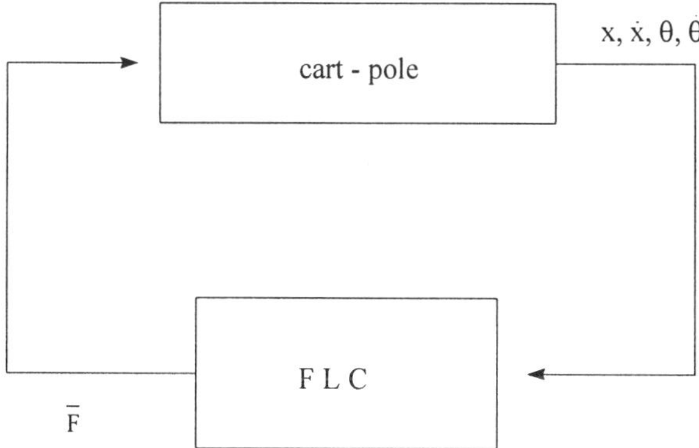

Figure 9.2 - The basic control loop for the cart-pole system is expanded dramatically. Adding adaptive capabilities to the controller requires additional components to identify when and to what extent the controlled system changes, and alter the controller based on these changes.

Development of the adaptive control system described in this chapter relied on a computer simulation of the problem environment. Such a computer simulation was described in Chapter 3 where the numerical integration of the governing equations of motion was discussed. The same computer simulation is used here but the system parameters can now be altered. For the reader's benefit, Table 9.1 lists both the system parameters and state variables associated with the cart-pole system.

Table 9.1 - System parameters and state variables for the cart-pole system

System parameters:

m	=	cart mass
l	=	pole length
m_c	=	mass of cart;
m_p	=	mass of pole;
l	=	length of pole;
μ_c	=	coefficient of friction of cart on track;
μ_p	=	coefficient of friction of pole on cart.

State variables:

x	=	cart position;
$\dot{x}$	=	cart velocity;
θ	=	angle of the pole with respect to the vertical;
$\dot{\theta}$	=	angular velocity of the pole.

The model equations are the basis of a mathematical model for the cart-pole system used to evaluate the effectiveness of potential fuzzy control strategies in Chapter 7. In the current chapter, this model is used for much more. The computer model is used by the analysis element to determine the magnitude of changes to the system parameters and by the learning element to investigate new and improved control strategies.

In the current effort, the control problem is made considerably more difficult by considering a system in which an unmeasured parameter of the system can change with time; a time-varying control problem. The parameter can change by a large amount. For instance the cart mass can increase by as much as 500 percent. These changes significantly alter the response of the cart-pole system to a given force stimulus. If the system parameters change too rapidly, we will not have time to determine the extent of the changes and update our control strategy accordingly.

9.3 ALTERNATIVE TO ADDITIONAL RULES

There are three main tasks that have to be accomplished by an adaptive, autonomous control system: (1) it must effectively manipulate the problem environment driving it from an arbitrary state toward a desired state, (2) it must recognize when the problem environment has changed in a manner that requires an adjustment to the control strategy (the rules and membership functions) being employed to accomplish the task of manipulating the problem environment, and (3) it must alter the control strategy in response to the new problem environment. A three-tiered computer software architecture has been developed to accomplish these three tasks. Together, GAs and fuzzy controllers include all of the capabilities necessary to produce powerful, efficient, and robust adaptive control systems and therefore form the core of the computer software architecture. To perform efficiently, such control systems require a ***control element*** to manipulate the problem environment, an ***analysis element*** to recognize changes in the problem environment, and an ***adaptive element*** to adjust to the changes in the problem environment. The remainder of this chapter describes the use of a GA in the development of a process control system that includes each of the three elements.

Figure 9.3 shows the adaptive process control system. The top two elements of Figure 9.3 are the elements of a traditional feedback control loop. The control element receives information from sensors in the problem environment concerning the status of the *condition variables*, i.e., x, $\dot{x}$, θ, and $\dot{\theta}$. It then computes a desirable state for a set of *action variables*, i.e., force to be added to the cart (F). Changes in the action variable forces the problem environment toward the setpoint ($x = \dot{x} = \theta = \dot{\theta} = 0$). This is the basic approach adopted for the design of virtually any closed loop control system. However, this basic loop includes no mechanism for adaptive control.

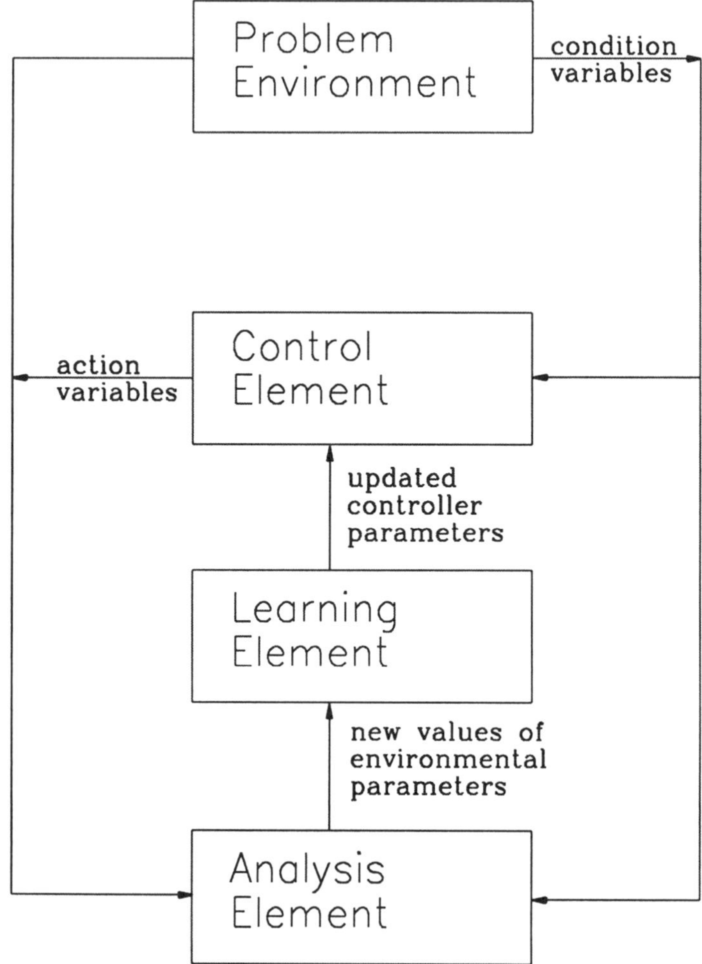

Figure 9.3 - The computer software architecture shown above allows for effective, robust, adaptive process control systems.

The adaptive capabilities of the system shown in Figure 9.3 are due to two new elements: (1) the analysis element and (2) the learning element. In general, the analysis element must recognize when a change in the problem environment has occurred. A "change," as it is used here, consists of any alteration of a system parameter that affects the response characteristics of the cart-pole system. Further, the change must occur to a parameter that is not included directly in the rule set. The analysis element uses information concerning the condition and action variables over some finite time period to recognize changes in the environment and to compute the new parameters appropriate to these changes.

Control Element

The control element receives feedback from the cart-pole system, and based on the current state of x, $\dot{x}$, θ, and, $\dot{\theta}$ must prescribe appropriate values of F. Any of a number of closed-loop controllers could be used for this element. However, since we have a perfectly good fuzzy controller for the cart-pole system, we use it here for the control element. The details of the cart-pole fuzzy controller are found in Chapters 3 and 7.

Analysis Element

The actions taken on the problem environment (input to the problem environment) and the subsequent changes to the problem environment (output of the problem environment) are evaluated using a computer model. If the model-predicted response of the problem environment matches the actual response of the problem environment (when the actual and model-simulated environments have been subjected to the same forcing function), then the system parameters are assumed to be unchanged. However, if the two environment responses do not match, then the parameters are considered to have changed.

The analysis element uses the computer model to evaluate the real-world system. It applies the same forces to a simulated cart-pole system that are being sent from the control element to the actual problem environment. Thus, the response of the computer model should be the same as the response of the real-world system. When it is not, then something in the real-world system has changed. Figure 9.4 shows a schematic of this concept. Note in the figure that at time t_c the response of the actual system shown on the left no longer matches the predicted response of the model as shown on the right.

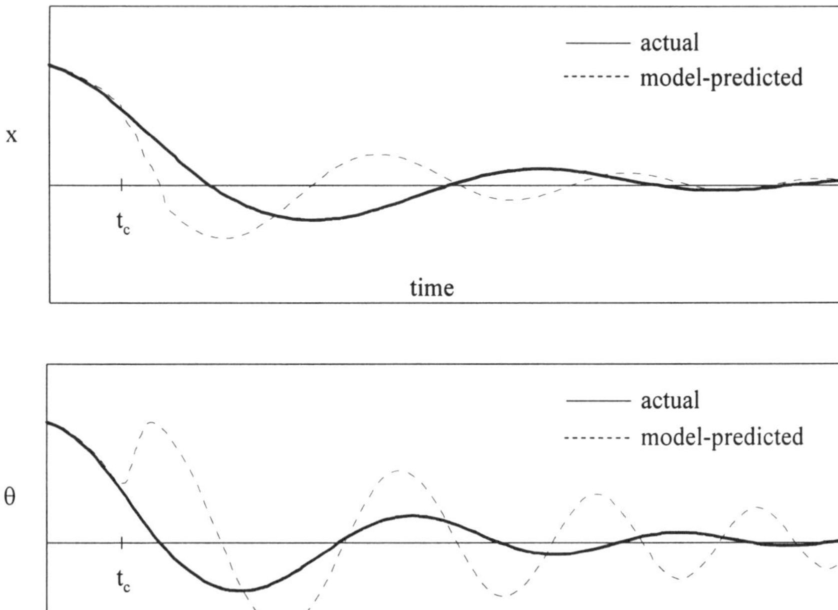

Figure 9.4 - When the response of the real-world system differs substantially from the response predicted by the analysis element's model simulation, it is assumed that the system parameters have changed. In the schematic above, this occurs at time t_c.

When the above approach is adopted, the problem of computing new system parameters becomes a curve fitting problem (Karr, 1991). The parameters associated with the computer model produce a unique response to a force, or action variables. Parameters must be selected by the analysis element so that the response of the model matches the response of the actual problem environment.

The GA used in the analysis element employs the familiar concatenated, mapped, unsigned binary coding scheme (Goldberg, 1989). The bit-strings produced by this coding strategy were of length 60: fifteen bits for each of the four system parameters that had to be represented, m, l, μ_c, and μ_p. The fitness function can be developed using the idea of curve fitting. Data exists describing the response of the real-world system. Thus, the GA must select the combination of system parameters that elicit this response for the given forcing function. Thus, the GA is looking for parameter values that will produce a series of states (x, $\dot{x}$, θ, and, $\dot{\theta}$) that match those in the real-world system.

This concept is by the following fitness function:

$$f = \sum_{i=0\,\text{sec}}^{i=30\,\text{sec}} \left(w_1 \left| x_i^{model} - x_i^{data} \right| + w_2 \left| \theta_i^{model} - \theta_i^{data} \right| \right)$$

where $w_1 = 1.0$ and $w_2 = 10.0$ are weighting constants that adjust for the fact that the values of x are generally an order of magnitude larger than the values of θ. With this definition of the fitness function, the problem becomes a minimization problem: the GA must minimize f, which, as it has been defined, represents the difference between the response predicted by the model and the response of the physical cart-pole system. When f (which is basically an error term) is driven to zero, then the response of the model exactly matches the response of the real-world system. Thus, it is assumed that the model is now working with an acceptable set of system parameters, a set that accurately reflects the real world.

Figure 9.5 compares the response of the cart-pole system being manipulated by the control element to the response of the cart-pole as simulated using parameters determined by the analysis element's GA. This figure shows that the responses of the two systems are virtually identical, thereby demonstrating the effectiveness of a GA in this application. The GA was able to locate the correct parameters after only 500 function evaluations, where a function evaluation consisted of simulating the cart-pole system for 10 seconds. Locating the correct parameters (the mass in the cart system was 2.0 kg; the GA computed a value of 1.992 kg) took approximately 10 seconds on a pentium-based personal computer.

It is important to note here that in the current situation, the "real-world" cart-pole system is itself a simulation. Therefore, we are able to stop time in the real-world system while we implement the analysis element. If we were working with a real cart-pole system, then the 10 seconds needed by the analysis element would be too long if the change in the system parameter is rapid; we could not figure out the magnitude of the changes, much less update our control strategies, before the pole would come crashing down. Industrial systems may mandate that a new control action be learned in less than 10 seconds. In such cases, the time the GA is allotted to update the model parameters can be restricted. However, there are numerous industrial systems in which the system parameters do not change rapidly relative to the requisite 10 seconds in the cart-pole system. It is these systems that are best suited to the adaptive control system described.

Once new parameters (and thus the new response characteristics of the problem environment) have been determined, the learning element must alter the control element. An updated control strategy must be developed that is effective at manipulating the new problem environment.

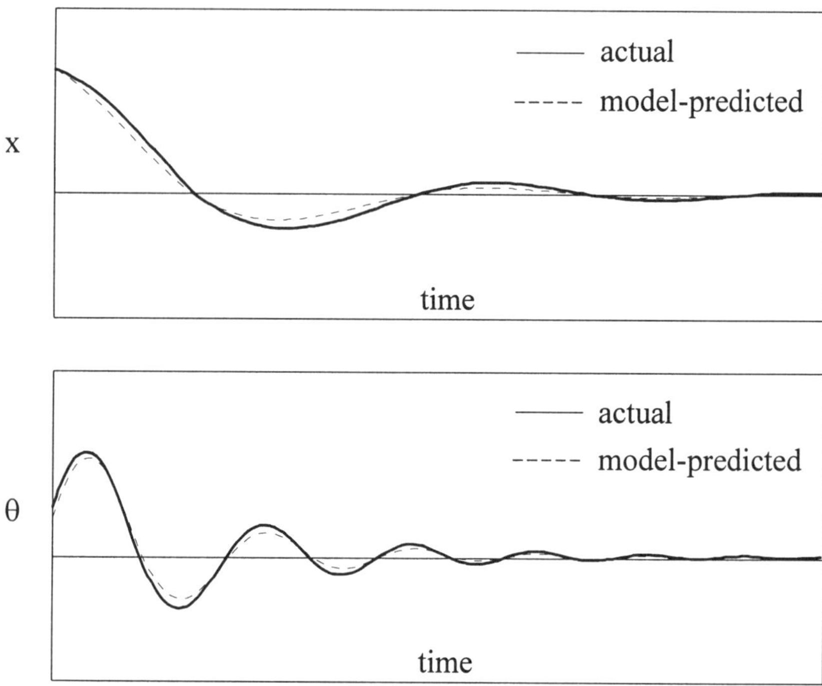

Figure 9.5 - The analysis element employs a GA to determine the correct values of system parameters. The fact that the response of the real-world system and the response of the simulated system are virtually identical indicates that the analysis element successfully accomplishes its assigned task.

Learning Element

The learning element alters the control element in response to changes in the problem environment. It does so by altering the membership functions (and in later chapters the rules) employed by the fuzzy controller. In this way, the control element is able to implement a new strategy that has been optimized for the new problem environment.

This task is equivalent to the fuzzy controller design problems presented in earlier chapters in which rules were discovered and membership functions were tuned to achieve a particular objective most effectively. The only difference is that the design is performed "on-line" (remember that for the cart-pole system, time is stopped while the analysis and learning elements accomplish their tasks) as opposed to the off-line designs performed earlier. The learning element has access to a computer model of the problem environment, it knows the current values of the system parameters in the real world, and it is also provided with the current state of the cart-pole system.

Altering the membership functions (the definition of the fuzzy terms in the rule set) is consistent with the way humans control complex systems. Quite often, the rules-of-thumb humans use to manipulate a problem environment remain the same despite even dramatic changes to that environment; only the meaning of the rules is altered.

The current adaptive control system uses a GA to alter the membership functions associated with a fuzzy controller, and this technique has been well documented. There is, however, one point that should be considered here. In the design of the cart-pole controller described in Chapter 7, the fitness function was developed to ensure that the controller could effectively drive the system to the desired setpoint from a variety of initial conditions. Here, the goal is to define membership functions that allow the controller to maintain the current state of the system. Thus, there is no need to look for membership functions that are effective from every possible initial condition case so the search problem is reduced dramatically. The fitness function employed allows for an increase in speed by a factor of four since it contains only one summation. The fitness function employed is:

$$f = \sum_{i=0}^{t_{\max}} w_1 (x - x_{desired})^2 + w_2 (\theta - \theta_{desired})^2$$

Although the additional summation did not matter in the preceding off-line design problems, the associated increase in speed is important in the adaptive control systems we are now considering.

In systems for which time is of the essence, the search problem can be reduced further. For instance, if the real-world cart-pole system is in one region of state space, the search can be constrained to those membership functions that have non-zero values in that particular region of the state space. In this way, the GA's task is much easier. Other than these two alternative considerations, the mechanics of the learning element are exactly as described in Chapter 7 and will not be repeated here.

9.4 RESULTS

An adaptive controller for the time-varying cart-pole system has been developed. To demonstrate its effectiveness, we use a particular situation in which the mass of the cart changes according to Figure 9.6. Notice here the mass is allowed to change dramatically, but not rapidly. Thus, this version of the cart-pole system presents a nice test bed for the adaptive controller.

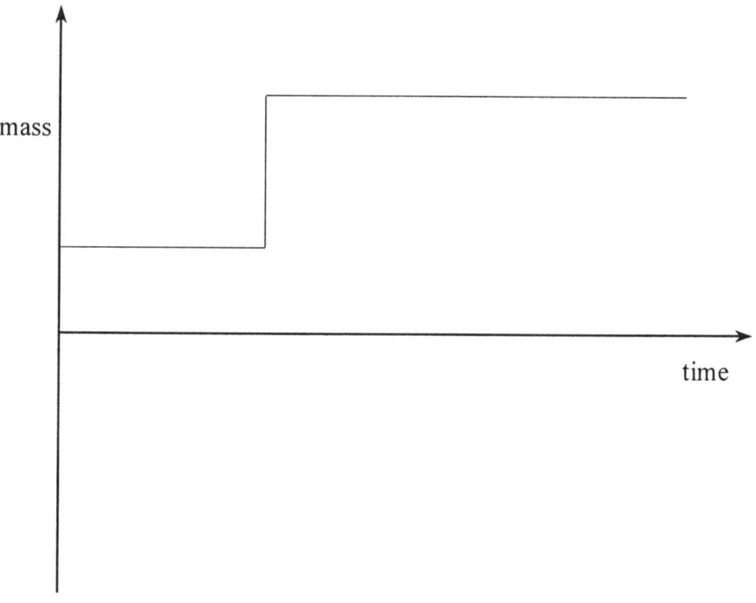

Figure 9.6 - The cart mass changes as a function of time. The magnitude of the mass changes dramatically, but the mass is not changing rapidly. Thus, this problem provides a nice test bed for the adaptive controller.

Figure 9.7 shows the effectiveness of the adaptive fuzzy control system that accounted for changes in the cart mass. This adaptive system was able to avoid the catastrophic failures of the pole falling over or the cart striking a wall despite dramatic changes in cart mass by altering its membership functions "on-line." Every time a change in cart mass was made, the analysis element recognized and quantified the changes. The mass changes were each identified to within 2 percent. The learning element employed a GA to locate new membership functions that were effective for the current state of the system. As can be seen in Figure 9.7, the adaptive GA-FC outperformed a non-adaptive FC that could not prevent the pole from falling over when the mass is altered.

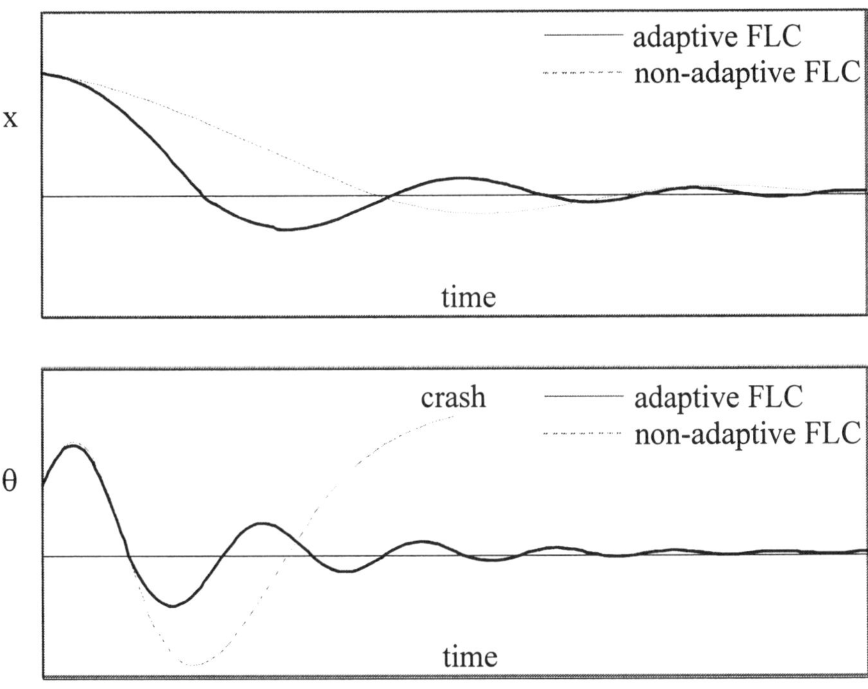

Figure 9.7 - The adaptive fuzzy controller is able to manage the changes in cart mass where the non-adaptive system holds firm to its rules and membership functions despite the fact that they are no longer valid for the altered system.

9.4 REVIEW/PREVIEW

This chapter introduced the concept of time-varying systems and the need for adaptive controllers. Specifically, a time-varying cart-pole system was introduced in which system parameters such as the cart mass, the pole length, and the coefficients of friction could change. The effect of changing a parameter was illustrated by changing the mass of the cart. An adaptive control system was developed to control such time varying systems. This adaptive control system included three elements: (1) a control element, (2) an analysis element, and (3) a learning element.

The control element consists of a fuzzy controller much like those described in previous chapters. It receives information about the current state of the system, and based on these values, prescribes an action to be taken on the problem environment. The analysis element recognizes when and to what extent the problem environment changes. It accomplishes this task by tracking the real-world system with a computer simulation. It then uses a GA to tune the parameters in a computer model until the response of the model matches the response of the real-world system. The learning element adjusts the control strategy being used by the control element based on information supplied by the analysis element. The learning element develops its updated strategy in the same way as the fuzzy systems were designed using GAs earlier in this book.

The adaptive control system described requires a computer model of the real-world system. For the cart-pole system, obtaining such a model was no problem because its physics lead to equations of motion which can be integrated numerically leading directly to a first principles model. However, there are systems for which acquiring or developing a computer model is a daunting task. Thus, we feel it is necessary to provide the reader with some approaches to computer modeling that we found to be effective. Thus, in the next section, we will introduce neural networks for modeling, further discuss the idea of using GAs for curve fitting, and present methods for data management.

The next section of the book deals with the development of computer models. However, it is important for the reader to keep in mind the software architecture presented for adaptive control presented in this chapter; it will be used in later applications we present.

REFERENCES

Goldberg, D. E. (1989). *Genetic algorithms in search, optimization, and machine learning*. Reading, MA: Addison-Wesley.

Karr, C. L. (1991). Genetic algorithms for fuzzy controllers, *AI Expert*, **6** (2), 26-31.

PART III
COMPUTER MODELING

Chapter 10

Genetic Algorithms for Model Tuning

10.1 INTRODUCTION

In Chapter 9 we introduced an approach to adaptive fuzzy control using genetic algorithms and a computer model of the system being controlled. For the cart-pole system, a reliable model was not difficult to obtain because the laws of physics led to a system of differential equations that were integrated numerically to produce an accurate model. However, for many systems this type of first-principle model is not available. For these systems, alternative modeling methods must be found. In this section of the book, we will consider several methods that can be used to develop various types of models including neural network models, fuzzy models, and regression models. The current chapter focuses on regression models and the use of GAs to tune such models in the presence of outliers in data. Outliers are points that are not consistent with the rest of the data set.

An empirical model is one alternative to a first-principle model. To construct empirical models, two things are necessary: (1) data representative of the real-world system and (2) a modeling equation. Ideally, the data should accurately depict the response of the system in the entire range of variables needed to control the system. The proper form of the empirical modeling equation must also be known. Generally, such an equation is selected from a search for an appropriate relationship such as a linear relationship, an exponential decay, or a power fit between variables (Mathews, 1992). For the appropriate equation its parameters can be adjusted so that the equation faithfully reproduces that data. Alternatively, an incomplete first-principle analysis can be supplemented with empirical constants that account for factors that are not fully understood.

The mineral processing industry, in particular, relies heavily on empirical or semi-empirical models. For models to accurately predict the response of a physical system, correct values of the model parameters must be obtained. The search for accurate values of empirical parameters can be a difficult, time-consuming task. Although the resulting models may not provide insight into the physics of the systems being modeled, they can be used just as effectively as first-principle models are used for training adaptive control systems.

The processes for determining values of empirical parameters for models are not always reliable and efficient. Classical statistical procedures such as least squares (LS) curve fitting have been used for data analysis across a broad spectrum of application domains. However, classical statistical techniques can be quite sensitive to *outliers*; the performance of many statistical methods degrades as erroneous points (outliers) appear in the data. A number of robust statistical techniques have been developed that attempt to overcome sensitivity to outliers (Huber, 1981). The virtue of these robust statistical methods is that small deviations caused by outliers in the data set hinders the models' performance only slightly and larger deviations do not cause the methods to fall apart altogether. Generally, these robust methods rely on iterative techniques where outliers are identified and delegated to a diminished role. Unfortunately, the problem of outlier detection can be computationally demanding.

Recently, a number of researchers have attempted to produce robust regression techniques that overcome the problem caused by outliers. Several effective techniques have been developed based on altering the LS approach to curve fitting. LS curve fitting consists of minimizing the sum of the squared residuals; a residual is the difference between the actual data value and the value predicted by the model. The least median squares (LMS) approach (Rousseeuw, 1984) replaces the sum of the squared residuals used in LS curve fitting with the median of the squared residuals, thereby resulting in a very different approach, an approach that is particularly effective at managing outliers. In LS curve fitting only one outlier can dramatically affect the accuracy of the result. However, in LMS curve fitting up to 50 pct of the data points can be outliers without degrading the accuracy of the curve fit.

Despite the effectiveness of the LMS method, it can face three situations that give it trouble: (1) when there is a large number of data points or when there is a large number of coefficients the method tends to be slow, (2) when speed is critical, the accuracy of the method suffers due to approximations that are made, and (3) nonlinear curve fitting problems are difficult for the LMS method (Rousseeuw and Leroy, 1987).

A promising approach to tackling the problems associated with performing regression on large data sets and with using nonlinear models is to combine the search capabilities of a GA with the curve fitting capabilities of LS. In fact, this approach has been shown to be effective in a number of problem domains (Karr, Stanley, and Scheiner, 1991) and was introduced in Chapter 5 of this book. Because of their rapid convergence characteristics, GAs can also be used to search for the constants needed in a LMS curve fitting problem. The GA exhibits properties that are desirable in the selection of LMS parameters. The LMS curve fit can often be achieved more rapidly using a GA, and the solution is more accurate where outliers are present.

In this chapter, a GA is used to implement LMS curve fitting methods. LMS curve fitting methods result in effective empirical computer models of several systems. The models presented in the current chapter include a hydrocyclone separator, a grinding system, and a column flotation circuit.

10.2 THE LS AND LMS METHODS

Empirical computer models can be constructed using *regression analysis*. The purpose of regression analysis is to fit equations of a very general standard form to observed variables. Although regression analysis is not employed in its pure sense in the models discussed in this chapter, the idea of using the least squares and least median squares regression criteria is employed.

The result of the modeling effort in this chapter is an empirical computer model that can then be used to perform interpolation and/or extrapolation on the system. The classical linear model assumes an equation of the form

$$y_i = x_{i1}\Theta_i + \ldots + x_{ip}\Theta_p + e_i \qquad (10.1)$$

for i = 1,...,N where N is the number of data points. The variables $x_{i1},\ldots,x_{ip}$ are the independent variables, and y_i is the dependent variable. In classical statistics, the error term, e_i, is assumed to be normally distributed with mean zero and unknown standard deviation, σ. One then estimates the vector of unknown parameters Θ from the N data points (x_i, y_i). Applying a regression estimator (which can be computed analytically for a number of model forms) to such a data set yields a set of estimated regression coefficients Θ'. Although the actual regression coefficients, Θ, are unknown, the independent variables can be multiplied by the estimated regression coefficients to obtain

$$y_{i'} = x_{i1}\Theta_{1'} + \ldots + x_{ip}\Theta_{p'} \qquad (10.2)$$

where y'_i is called the predicted value of y_i. The residual r_i of the ith case is the difference between what is actually observed and what is predicted:

$$r_i = y_i - y_{i'}. \qquad (10.3)$$

LS curve fitting is the most popular regression approach and dates back to Gauss and Legendre (see Plackett, 1972 for some historical discussion). In the LS method, the sum of the squares of the residuals is minimized. Mathematically, the objective is to minimize

$$\sum_{i=1}^{N} r_i^2. \qquad (10.4)$$

The idea is to optimize the fit by making the sum of the residuals as small as possible. The method's popularity is easily understood when one realizes that at the time of the method's conception there were no computers, and the fact that the estimated regression coefficients could be calculated explicitly via some matrix algebra, made LS the most practical approach to curve fitting. Even today, most statistical packages utilize the LS method because of tradition and speed of computation. Actually, there are a number of situations for which LS curve fitting is more than adequate. However, there is a situation for which the method is not suitable.

A major weakness of the LS method is apparent when the data set that is being fitted contains outliers. These data points, which for one reason or another do not fit the assumed form of the model, can cause the LS method to give bad results. A single outlier can dramatically affect the results of the LS method. Generally, the number of outliers it takes to contaminate a method is defined by a breakdown point. The *breakdown point* is the largest fraction of contamination that can cause the estimator to take on values arbitrarily far from the assumed model. For the LS method, the breakdown point tends to zero, which means that only one outlier is needed to cause the method to give inappropriate results. Under tightly controlled conditions (or when some pre-screening is performed), the number of outliers that are present in a data set can be minimized if the data points have at most three dimensions (Rousseeuw and Leroy, 1987). However, in a number of real-time engineering applications, data sets cannot be adequately purged of outliers. Fortunately, there is a regression technique that is capable of handling the occurrence of outliers more efficiently than the LS method.

The *LMS method of regression* minimizes the median squared residual. An estimator is used to compute each of the N residuals and the median residual is found. It is this median residual that is minimized. The LMS method has a breakdown point of 50 pct since the points are sorted and the poorest half of the points do not influence the estimator. As it turns out, this estimator is very robust with respect to outliers. However, the method is not without its problems.

When the data set is large, the method used at present to develop an LMS model can take a long time to converge. This occurs because the algorithm is based on examining the possible combinations of pairs of the data points, and for each combination a model is computed that includes performing a computer sort. There are situations wherein it is infeasible to examine all of the possible combinations. In such instances the number of possibilities considered must be reduced in some meaningful way. However, the savings gained in computational time can be offset by a failure to locate the best possible solution. Additionally, this approach to LMS regression is limited by the fact that the modeling equation must pass through two of the data points exactly in high order equations. There are, however, situations in which the best fit does not pass directly through any of the data points.

The remainder of this chapter focuses on using a GA to locate the estimator for a LMS regression; here estimator refers to the constants in the chosen model equation. Since we have addressed the requirements of the GA in several places previously, the details of the application are summarized. Results are provided that demonstrate the effectiveness of using GA LMS for developing empirical computer models. Most of the examples come from the mineral processing industry, but the approach is applicable to many engineering problems.

10.3 HYDROCYCLONE SEPARATOR

The use of hydrocyclones is the standard method in the mineral processing industry of separating the particles in a slurry by size or by density. Hydrocyclones are continuously operating separating devices that utilize centrifugal forces to accelerate the settling rate of particles. They have achieved wide-spread use because of their simplicity, their durability, and their relatively low cost.

A typical hydrocyclone is shown in Figure 10.1. The lower portion of a hydrocyclone is a conical vessel with an opening at the bottom to allow for the removal of the coarse or heavier particles. The conical section is joined to a cylindrical section, the top of which is closed with the exception of an overflow pipe known as a *vortex finder*. The vortex finder prevents the feed material that enters the cylindrical section tangentially from going directly into the overflow. Separation occurs in the cylindrical section due to the existence of a complex velocity distribution that carries the coarse particles to the bottom and the fine particles out the top.

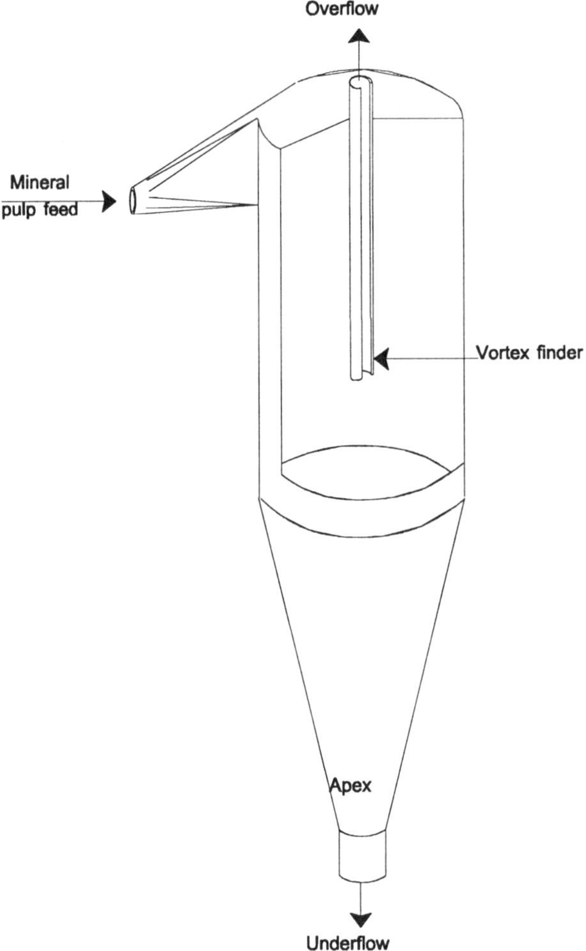

Figure 10.1 - Hydrocyclones have become the standard separation device in a number of industries including both the mineral processing industry and the separations industry.

Plitt (1976) presented an empirical model to predict the d_{50} or *split size* that is still used extensively today. The split size is that size particle (given by diameter of the particle) that has an equal chance of exiting the hydrocyclone either through the underflow or the overflow, and is often used to quantify the efficiency of this separation process. Plitt's model has the following functional relationship:

$$d_{50} = \frac{C_1 D_c^{C_2} D_i^{C_3} D_o^{C_4} \exp[C_5 \phi]}{D_u^{C_6} h^{C_7} Q^{C_8} \rho^{C_9}} \tag{10.5}$$

where D_c is the diameter of the hydrocyclone, D_i is the diameter of the slurry input, D_o is the diameter of the overflow, D_u is the diameter of the underflow, h is the height of the hydrocyclone, Q is the volumetric flow rate into the hydrocyclone, ϕ is the percent solids in the slurry input, and ρ is the density of the solids. The empirical constants, C_i, are selected so that the model accurately matches data that has been collected from actual hydrocyclone separators.

A number of approaches have been developed for determining the empirical constants associated with this particular model of a hydrocyclone. Most of the approaches involve the use of statistical routines, each of which has some potential drawbacks (Karr et al, 1995), many of which can be overcome by using a GA. The problem at hand is similar to the curve-fitting problem of Chapter 5. The coding scheme that was employed in the Ree-Eyring equation example of Chapter 5 is again used here. The fitness function of Chapter 5 is similar, except that the LS criteria has been replaced with the LMS criteria resulting in the following fitness function:

$$f = med\,[(y_{i'} - y_i)^2] \tag{10.6}$$

where f is the fitness function value, *med* is the median value, y is an actual data value, and y' is an estimated data value using the parameters defined by a particular GA string.

A GA was used to select the values of C_1 through C_9 used in the empirical model of a hydrocyclone as shown in equation (10.1). Figure 10.2 demonstrates the effectiveness of using a GA for tuning the empirical constants. In this plot, the actual d_{50} size is plotted against the d_{50} size predicted using the model with the empirical constants as selected by a genetic algorithm. Note that the model would exactly reproduce the data if all of the points shown in Figure 10.2 fell on a line of 45°. Although the GA tuned model fits the data well for most of the data shown (excluding the outliers in the lower range of d_{50}) it systematically departs from the 45° line at high values of d_{50}. This modeling error indicates that either the GA has not determined the exact values of the coefficients or the mathematical relationship upon which the model is based is inadequate.

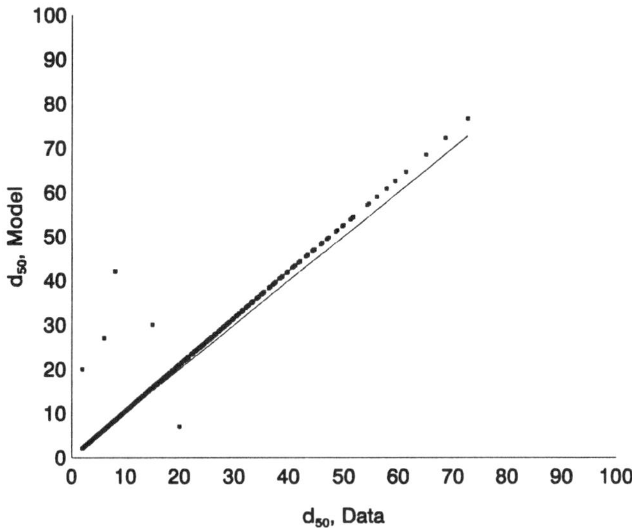

Figure 10.2 - The GA tuned empirical model of the hydrocyclone separator matches the data obtained for real-world hydrocyclones well.

It is important to note that the GA determined the empirical constants in roughly 10 minutes, whereas a statistical package took roughly 9 hours to locate constants that provided the same quality fit to the data. This time of computation becomes critical when the tuning of an empirical model is being done as a part of a real-time, adaptive process control system such as the one discussed in Chapter 9.

10.4 GRINDING

Grinding is a necessary component in the processing of a number of ores. Optimizing the grinding process, however, is a difficult task requiring innovative techniques and accurate computer models. But, because of the high energy costs and the large amount of ore processed, improvements in the efficiency of grinding have a dramatic economic impact.

The grinding process is characterized by several performance measures, all of which are important in various circumstances. Generally, there are four measures that are especially important indicators of grinding efficiency: (1) *fineness* of the ground product, (2) *energy* costs associated with the process, (3) a *viscosity coefficient*, and (4) a *viscosity exponent*. Mehta, Kumar, and Schultz (1982) have developed an empirical model of grinding that accurately mirrors the response of a coal grinding process. In that work, they studied the ability of a simple genetic algorithm to optimize the performance of their computer model of a particular grinding circuit.

A genetic algorithm was used to tune the computer model of a grinding process. Only two of the four indicators of grinding efficiency are considered: fineness and energy consumption. However, the steps used to tune the models that predict fineness and energy consumption could also be used to tune models that predict viscosity characteristics. The pertinent modeling equations are:

$$\begin{aligned} F = {} & C_1 + C_2\, x_s - C_3\, x_B + C_4\, x_D + C_5\, x_M \\ & - C_6\, x_s^2 - C_7\, x_B^2 + C_8\, x_D^2 - C_9\, x_M^2 \\ & + C_{10}\, x_{sB} + C_{11}\, x_{sD} - C_{12}\, x_{sM} \\ & + C_{13}\, x_{BD} - C_{14}\, x_{BM} - C_{15}\, x_{DM} \end{aligned} \tag{10.7}$$

and

$$\begin{aligned} E = {} & C_{16} + C_{17}\, x_s - C_{18}\, x_B + C_{19}\, x_D + C_{20}\, x_M \\ & - C_{21}\, x_s^2 - C_{22}\, x_B^2 + C_{23}\, x_D^2 - C_{24}\, x_M^2 \\ & + C_{25}\, x_{sB} + C_{26}\, x_{sD} - C_{27}\, x_{sM} \\ & + C_{28}\, x_{BD} - C_{29}\, x_{BM} - C_{30}\, x_{DM} \end{aligned} \tag{10.8}$$

where F is the weight percent of material less than 38 millimeters, E is the energy in kilowatt hours per ton, x_s is the percent solids by weight, x_B is the maximum ball size, x_M is mill speed, x_D is dispersant addition, x_{ij} represents the product of x_i and x_j, and C_1 through C_{30} are the empirical constants selected so that the model equations accurately reproduce data obtained from a physical system. It is important to note that C_1 through C_{15} can be determined entirely independent of C_{16} through C_{30}. Thus, there are two separate 15 parameter search problems to be solved.

Development of an efficient empirical computer model of a grinding circuit actually depends on tuning 15 parameters for each of four separate model equations (only two such equations are considered here). The very same approach used in the previous sections of employing a genetic algorithm for tuning computer models can be used here. The coding scheme outlined in the hydrocyclone section can easily be used, as can the fitness function definition. Thus, all that is left is to present some results.

Figures 10.3 and 10.4 demonstrate the effectiveness of using a GA for tuning the empirical constants associated with the grinding models. In these plots, the fineness and energy data are plotted against the computer-predicted values. These figures demonstrate the ability of a GA to determine empirical constants for the grinding models.

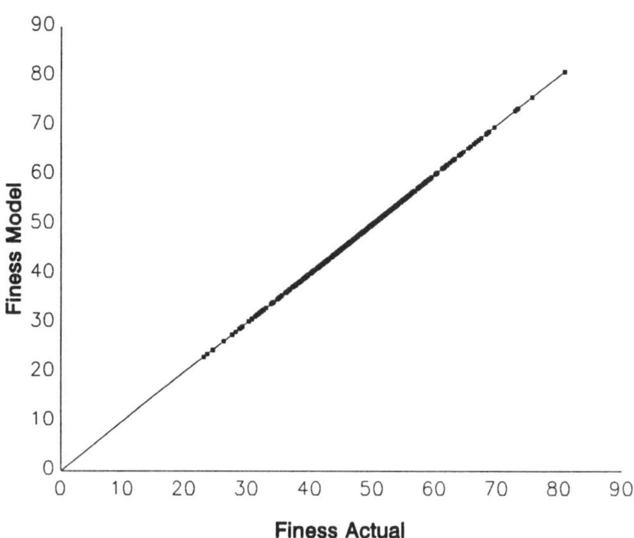

Figure 10.3 - A GA effectively tuned the fineness model of grinding as indicated in the figure in which data values are plotted against model-predicted values.

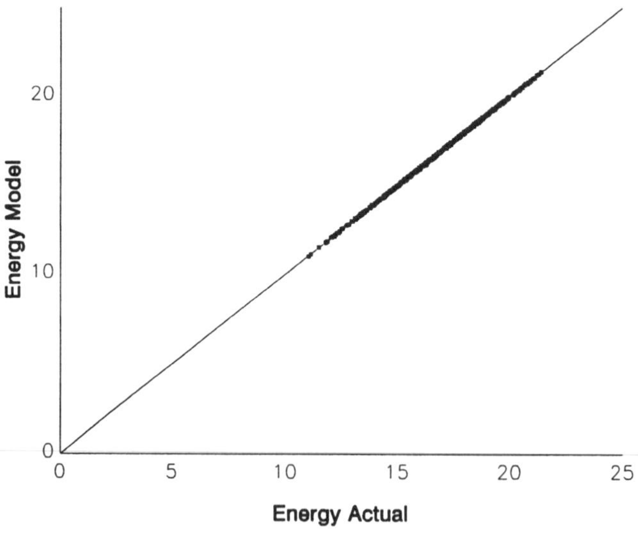

Figure 10.4 - A GA effectively tuned the energy model. The plot above indicates that the empirical model accurately matched the data values over a range of values.

10.5 COLUMN FLOTATION CIRCUIT

Flotation is an established method for concentrating minerals (Sastry, 1978). Separation occurs when selected particles are rendered hydrophobic (water repellent) with chemical agents (called collectors) and are attached to air bubbles added to a vigorously stirred flotation cell containing a mineral-water slurry. The bubbles and their attached mineral particles rise to the top of the flotation cell to form a froth. The froth is collected and the mineral recovered. Such conventional flotation cells have been a cornerstone of the mineral processing industry for a number of years. However, column flotation that occurs in a taller unstirred flotation cell is becoming increasingly popular as an alternative to conventional flotation cells because of its efficiency and mechanical simplicity. Several copper-molybdenum operations in the United States and Canada are currently operating flotation columns while numerous other ventures are running pilot-plant tests (Agar, Huls, and Hyma, 1991). The popularity of these separators should continue to grow as a better understanding of the underlying mechanics of the operation is gained.

Figure 10.5 shows a schematic of a column flotation unit. The chemically treated mineral slurry (feed) is fed into the column approximately one-third of the way down from the top of the column. Bubbles are introduced at the bottom of the column. As the feed particles sink and the bubbles rise, collisions occur. Due to the chemical treatment of the feed, some particle species attach to the bubbles during collision and rise to the top to form a foam or froth zone, while others are repelled by the air thereby remaining unattached and sink to the bottom. Thus, a separation occurs.

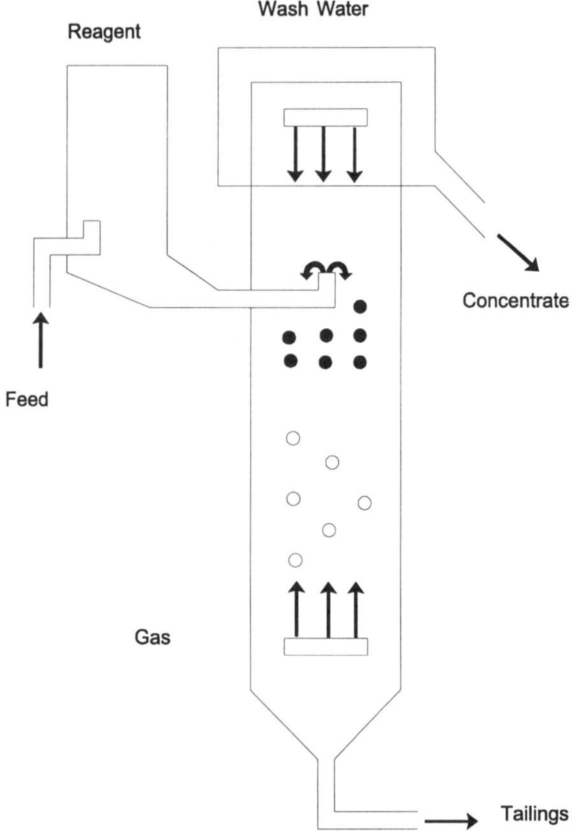

Figure 10.5 - Column flotation units are becoming increasingly popular separators because of their mechanical simplicity.

Because of the increased use of column flotation units, researcher have begun to investigate their mathematical underpinnings. The result of these efforts has been the development of systems of mathematical equations that describe column flotation, or separate events associated with column flotation. Some of the more noteworthy research has been reported by Finch and Doby (1990), Herbst and Rajamani (1989), Luttrell, Adel, and Yoon (1987), and Ynchausti, Herbst, and Hales (1988). In the current effort, the Finch and Doby model has been used because it has been well documented, is relatively straightforward to program, and provides accurate results.

The Finch and Doby model of column flotation consists of a number of empirical and semi-empirical equations that model individual aspects of flotation separately. Each of the equations has parameters that must be adjusted to fit conditions in a specific flotation cell. Many of these equations are concerned with calculating the number of collisions between the bubbles and the particles. These equations incorporate factors such as particle and bubble size and concentration, and the relative velocity between particles and bubbles. A very important equation is used to calculate the time it takes after collision for a particle to slide around the bubble. The net result of these various calculations is a rate constant for each species in the flotation cell. The Finch and Doby model of column flotation also relies heavily on a property of hydrophobic minerals called *induction time*. Induction time is the time required, once a hydrophobic particle has collided with a bubble, for the particle to attach to the bubble. If the induction time is less than the sliding time the bubble and particle become attached. Once the attachment has occurred, the particle will usually remain attached to the bubble until the bubble/particle aggregate rises to the liquid/froth interface level. Also, it is in practice often a good assumption that an ore particle that reaches the interface level will remain attached to its bubble and be carried out with the concentrate. The foam is sprayed with water to wash undesired and non hydrophobic particles back into the flotation cell.

Often the material concentrated by flotation will only be partially purified. If this is the case it must be re-floated in additional cells. In addition, some of the ore mineral will remain unfloated and this material must also be re-floated. The resulting assembly of flotation cells is called a flotation circuit. The exact amount of mineral removed by each flotation cell is determined by the residence time for a species in the flotation cell. The residence time is determined by a vessel dispersion number that is a measure of the degree of mixing in the vessel.

Data on the column circuit considered in this effort was presented in the thesis by Alford (1991). Alford's data is taken from a circuit used in an industrial application at the Mount Isa Mines Limited installation in Australia.

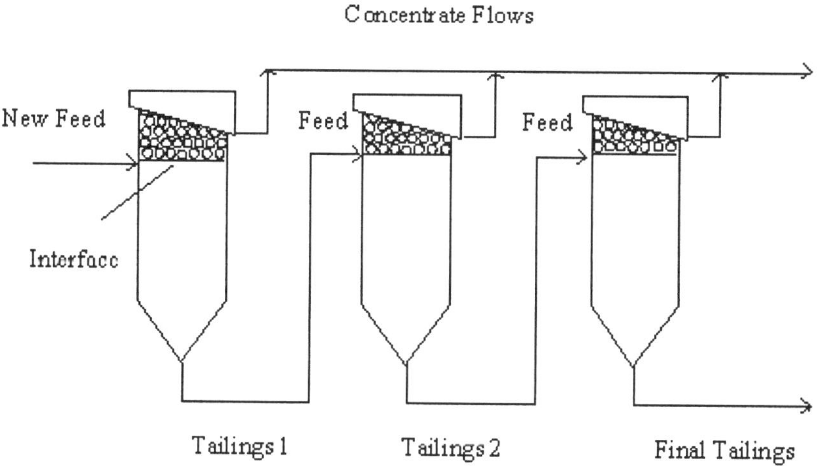

Figure 10.6 - Column flotation units are used effectively in series as shown in the schematic above.

A GA has been used to tune selected parameters spread throughout a number of individual equations associated with flotation events in each column in order to make the values predicted by the model match as nearly as possible the behavior of the actual circuit as recorded by the data in Alford's thesis. In addition it was necessary to determine a value for the vessel dispersion number. Since vessel dispersion number is largely a function of the size and shape of the column and since the physical dimensions of the three columns were identical, the three were assumed to have the same vessel dispersion number. Since induction time is a chemical property, the sphalerite and galena induction times were also assumed to be the same for each of the three columns.

The genetic algorithm was set up to minimize the mean squared error between the balanced data values based on observed flows in the Mount Isa Mines circuit and the corresponding values obtained from the computer model. There were 30 data values used in computing the mean squared error, and these data values included information about the percent solids, the percentage of lead, the percentage of zinc, the percentage of iron, and the flow volume in each of the six flows produced by the circuit. The final mean squared error attained by the genetic algorithm was 0.0003. The precise values achieved by the model are listed in Table 10.1 along with the balanced values based on observed data for comparison. (The percent solids column is omitted.) Each table entry is of the form V_m/V_b, where V_m is the value produced by the computer model and V_b is the balanced value from Alford's data.

Table 10.1 – The comparison of values below demonstrates the effectiveness of the genetic algorithm in effectively tuning the computer model of the flotation circuit.

Flow Name	% Lead	% Zinc	% Iron	Flow Volume
Stage 1 Concentrate	11.7/11.8	41.0/40.8	12.4/12.5	12517.8/12632.1
Stage 1 Tails	5.7/5.7	14.6/14.5	26.6/26.7	31942.1/31828.1
Stage 2 Concentrate	11.1/10.7	36.5/37.4	14.0/14.2	4597.7/4457.6
Stage 2 Tails	5.0/5.0	11.6/11.7	28.3/28.2	39344.5/39370.4
Stage 3 Concentrate	11.1/11.1	34.3/32.5	17.5/16.9	2684.4/2710.0
Stage 3 Tails	4.5/4.5	9.7/9.8	29.2/29.2	48660.1/48660.5

10.6 REVIEW/PREVIEW

Empirical computer models play a key role in a variety of industries for tackling both design and control problems. Generally, the calibration or tuning of these models, the selection of values for the empirical constants, is a time-consuming task. Fortunately, the search capabilities of genetic algorithms can be used to help solve the problem of selecting these constants.

There are numerous criteria for judging accuracy of empirical models. The traditional least squares criteria is in widespread use, but it does not perform well in the presence of outliers. Thus, a more robust technique, least median squares, has been presented. The LMS method works well when used in conjunction with genetic algorithms to produce regression-based computer models based on data obtained from physical systems. The effectiveness of this approach to LMS computer modeling has been demonstrated on several systems including a hydrocyclone, a grinding circuit, and a column flotation circuit.

One of the major drawbacks of empirical models is that the form of a modeling equation must be known in advance. Such an equation is not always available, and thus alternative techniques must be considered. Frequently, data collected from a plant environment is the only information available to the modeler. In such a situation, regression techniques are extremely difficult to develop, as an effective equation form must be discovered. Alternatively, the modeler can employ techniques that are capable of modeling problems using only the available data. Such a technique is presented in the next chapter. Neural networks are computational paradigms of the human brain and their associated theory has been developed by researchers from the field of artificial intelligence. As will be seen in the following chapter, neural networks can be used to develop accurate computer models.

REFERENCES

Agar, G. E., Huls, B. J., & Hyma, D. B. (Eds.) (1991). *Column '91.* The Canadian Institute of Mining and Metallurgy and Petroleum.

Alford, R. A. *Modelling and design of flotation column circuits.* Doctoral dissertation, Queensland, Australia: The University of Queensland, 1991.

Finch, J. A., and Doby, G. S. *Column flotation.* Toronto: Pergamon Press, 1990.

Herbst, J. A., and Rajamani, K. Models for the dynamic optimization of mineral processing plant performance. In K. V. S. Sastry and M. C. Fuerstenau (Eds.) *Challenges in mineral processing.* SME, 709-739, 1989.

Huber, P. J. (1981). *Robust statistics.* New York, NY: John Wiley & Sons.

Karr, C. L, Weck, B., Massart, D. L., & Vankeerberghen, P. (1995). Least median squares curve fitting using a genetic algorithm. *Engineering Applications of Artificial Intelligence*, **8**(2), 177-189.

Karr, C. L. Optimization of a computer model of a grinding process using genetic algorithms. In H. El-Shall, B. Moudgil, and R. Weigel (Eds.), *Beneficiation of Phosphate: Theory and Practice* (pp. 339-345). Littleton, CO: Society for Mining, Metallurgy, and Exploration, Inc., 1993.

Karr, C. L., Stanley, D. A., & Scheiner, B. J. (1991). *A genetic algorithm applied to least squares curve fitting* (Report of Investigations number 9339). Washington, DC: U.S. Department of the Interior, Bureau of Mines.

Luttrell, G. H., Adel, G. T., and Yoon, R. H. *Modeling of column flotation.* SME Annual Meeting, Preprint number 87-130, 1987.

Mathews, J. H. (1992). *Numerical methods for mathematics, science, and engineering.* Englewood Cliffs, NJ: Prentice Hall.

Mehta, R. K., Kumar, K. K., and Schultz, C. W. Multiple objective optimization of a coal grinding process via simple genetic algorithm, Society for Mining, Metallurgy, and Exploration, Inc., Littleton, CO, preprint number 92-108, 1982.

Plackett, R. L. (1972). Studies in the history of probability and statistics XXIX: The discovery of the method of least squares, *Biometrika*, **59**, 239-251.

Plitt, L. R. A mathematical model of the hydrocyclone classifier. *CIM Bulletin*, **69**: 114-123, 1976.

Rousseeuw, P. J. (1984). Least median of squares regression, *Journal of the American Statistical Association*, **79**, 871-880.

Rousseeuw, P. J., and Leroy, A. M. (1987). *Robust regression and outlier detection.* New York: John Wiley & Sons.

Sastry, K. V. S. (1978). *Beneficiation of mineral fines, problems and research needs*. Report of a workshop held at Sterling Forest, New York.

Ynchausti, R. A., Herbst, J. A., and Hales, L. B. Unique problems and opportunities associated with automation of column flotation cells. *Column Flotation '88*, SME, 27-33, 1988.

Chapter 11

Neural Networks for Computer Modeling

11.1 INTRODUCTION

In the previous chapter, we saw that a genetic algorithm could be used to develop computer models by adjusting empirical parameters in modeling equations. The approach was based on least-median-squares curve-fitting and proved to be robust because the least-median-squares criteria rejected outliers in the data. However, there are numerous occasions for which a modeling equation is not available. In these instances the approach presented in Chapter 10 is not possible. Although there are several approaches to computer modeling, we again turn to the field of artificial intelligence, which offers a very general modeling tool that does not require specific modeling equations, neural networks (NNs). NNs have been used as models in a number of industries (Wasserman, 1989; Pao, 1989). They are arguably the most flexible modeling technique known.

Neural networks are computational systems composed of nodes (sometimes called neurons, neurodes, or units) and the connections between these nodes. Neural computing attempts to emulate the functions of the mammalian brain, thus mimicking thought processes. NNs differ from most computer algorithms in that they are not "programmed" rather they are "trained." The training uses data obtained from the system being modeled. The NN processes the data and adjusts the weighted connections between the nodes so that the network can, as accurately as possible, reproduce the training data. However, the NN does far more than simply memorize the data it is presented. Instead, it uses the training set to develop a relationship between the input and output parameters. Information concerning these relationships is contained in the weighted connections (appearing as weight matrices in actual implementation of a NN) of the NN. In this way, NNs are able to extend the knowledge they contain to data sets for which they have not been trained. Thus, they are powerful modelers requiring no prior knowledge of the system being modeled other than the number of input and output parameters.

In the current chapter we will introduce the mechanics of an extremely popular NN paradigm, a backpropagation NN algorithm (McClelland and Rumelhart, 1988). This algorithm is then applied to modeling three systems from the mineral processing industry: (1) a grinding circuit, (2) a hydrocyclone, and (3) a flotation circuit.

10.2 NEURAL NETWORK BASICS

There is a wide variety of neural network architectures (the arrangement and connection of neurons, and training algorithms) that have been used successfully to solve numerous problems. However, despite the differences in the architectures, NNs are based on a common component, the neuron that is the basic processor in NNs. In their most basic form, neurons are computational units that receive multiple signals, sum these signals, and compute an output based on an activation function (generally nonlinear) that is a function of the sum of the signals received. Each neuron has one output, which is generally related to the state of the neuron (its activation). The output of each individual neuron is passed on to (potentially numerous) other neurons in various positions in the network via weighted connections. Each neuron receives inputs over these connections, called synapses, from other neurons. The inputs are the "activations" of the incoming neurons multiplied by the weights of the synapses along which they travel. The activation of the neuron is computed by applying a threshold function to this product. The threshold function is used to introduce nonlinearities into the neural network. An abstract model of the basic neuron is shown in Figure 11.1.

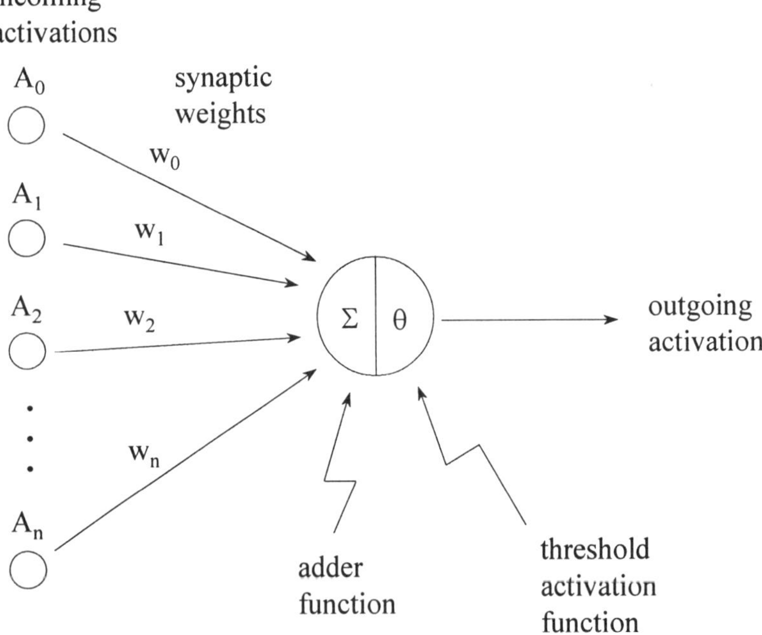

Figure 11.1 - The basic neuron receives input in the form of activations from other neurons, processes the information, and provides its own outgoing activation that is then sent to other neurons in the NN.

The activation or threshold function, f(x), is a nonlinear function. Nonlinearity is an extremely important attribute to NNs because it allows them to model complex relationships between input and output parameters. The most commonly used nonlinear function that is appropriate for NNs is the sigmoidal function defined by

$$f(x) = \frac{1}{1 + e^{-x}} \tag{11.1}$$

and shown in Figure 11.2.

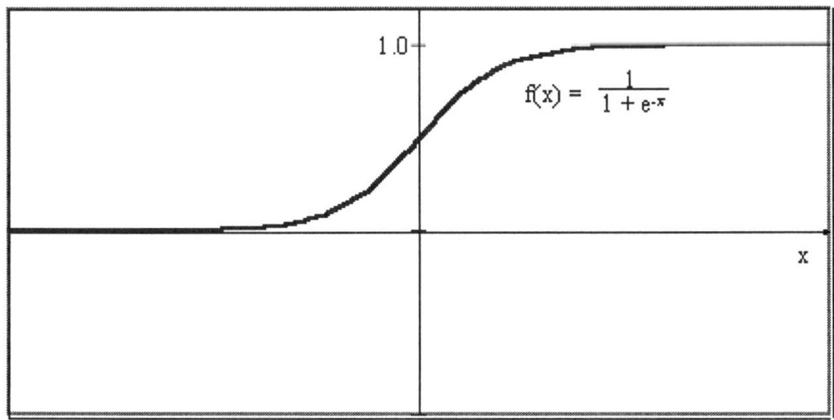

Figure 11.2 - Sigmoidal functions are nonlinear. They provide NNs with the capability of representing nonlinear relationships such as the ones commonly needed in computer modeling efforts.

The individual neurons of neural networks are placed in layers. The most popular configuration is to have a three-layer network. The neurons in the layers are interconnected by weighted synapses. Figure 11.3 shows a schematic of a typical backpropagation network consisting of three layers: (1) an input layer, (2) a hidden or middle layer, and (3) an output layer. The input layer neurons receive an activation signal from the physical environment (data from the system to be modeled). These input layer neurons then sum the input signals and compute an activation based on their activation or sigmoidal functions. The resulting activation, or output, is passed to each of the neurons in the hidden layer (for a fully connected network each neuron in the input layer is connected to each neuron in the hidden layer, and each neuron in the hidden layer is connected to each neuron in the output layer). Along the way, the activation signals are multiplied by the weights associated with each of the synaptic connections along which the signals are passed. This is equivalent to taking a weighted average of the activations from the contributing neurons. These weighted activation signals serve as inputs to the hidden layer neurons. And, selecting the weights in the network such that it gives the correct answers is the whole problem.

The hidden layer neurons perform calculations with the signals similar to the calculations performed by the input layer neurons resulting in activation signals that are passed to the output layer neurons. The output neurons use these weighted input activation signals coming from the hidden layer neurons to compute their output activation signals. These activation signals coming from the output neurons are the output of the neural network. This process by which an input signal (a vector) is received from the physical world, multiplied by weights associated with the connections between input and middle layer neurons (these weights are actually matrices in NN implementation), nonlinearized with activation functions, multiplied by a second set of weights associated with the connections between middle and output layers, and nonlinearized again with activation functions, is called a forward pass. This pass is used both in training an NN and in using a trained NN to model a system. There is no improvement or learning present in this forward pass. It is, however, in a backward pass that an NN adjusts its weights so that it accurately models the training sets.

The backward pass through a neural network involves using derivative information to adjust weight values so that the network accurately models the response of the system being modeled. The derivatives are in the form of changes in error with respect to changes in weight values. This derivative information can be used to propagate the error back through the network in the form of adjustments to the weight values. The result is a derivative-based search that produces a well-trained network capable of predicting the response of a variety of systems.

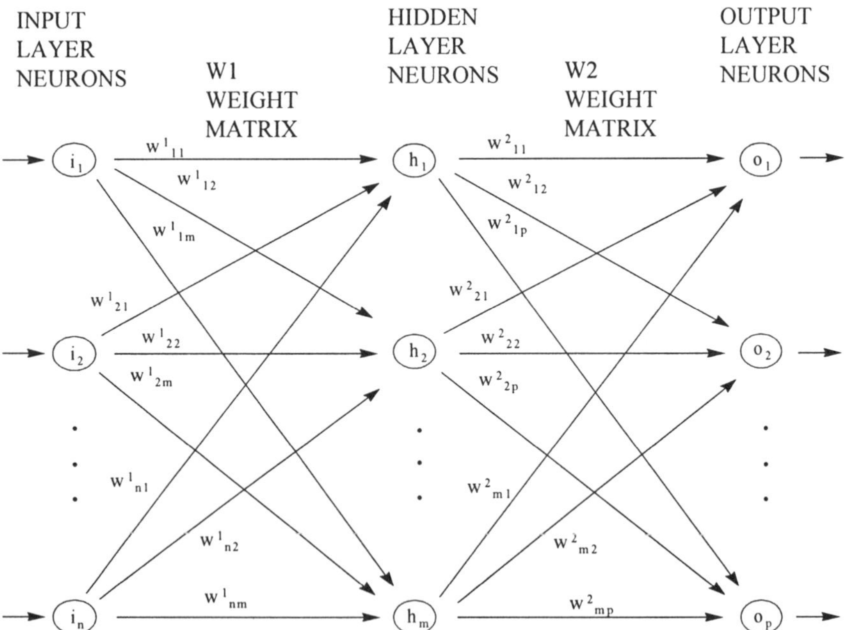

Figure 11.3 - Typical backpropagation NNs consist of three layers of neurons interconnected by weighted synapses.

We now have provided an architecture in which numerous neurons are connected into layers. The network has a mechanism for accepting an input signal and converting it into an output signal. The relationship between the network's output signal and its input signal is determined by the weighted connections that are adjusted in a backward propagation of errors. The backward propagation of errors through the network represents the training phase of an NN while the forward pass is the actual implementation of the NN. Generally, the implementation is easy and fast while the training can be complex and time-consuming.

Before we discuss this process, it is prudent for a moment to consider what the input and output signals for a modeling problem might be. As an example, recall the hydrocyclone model presented in Chapter 10. In this modeling problem the goal was to model the relationship between several system parameters. Specifically, the goal was to model the relationship

$$d_{50} = f(D_c, D_i, D_o, \phi, D_u, h, Q, \rho) \tag{10.5}$$

where D_c is the diameter of the hydrocyclone, D_i is the diameter of the slurry input, D_o is the diameter of the overflow, D_u is the diameter of the underflow, h is the height of the hydrocyclone, Q is the volumetric flow rate into the hydrocyclone, ϕ is the percent solids in the slurry input, and ρ is the density of the solids. Thus, an input data set used to train an NN would contain measured values of each of the parameters on the right-hand side of the above equation. A corresponding value of d_{50} would also be available to the NN so that it could determine the accuracy of its computed value of d_{50}. This comparison is used to compute errors that are propagated back through the network in the backward pass.

Backpropagation NNs take a training example, compute an output signal, and then compute an associated error based on the known solution. Initially, errors occur mainly because the weighting values on the synapses are not accurate. Thus, a mechanism is needed to adjust the weights so that the network "will get a better answer." Weight adjustment by the backpropagation of errors is a relatively generic concept. There are many, many ways to do this. However, the classic backpropagation algorithm uses a derivative based approach. Perhaps the most effective way to present the algorithm is to provide a step-by-step procedure for implementing and training an NN.

Before presenting the algorithm, let's review a couple of key issues. First, the training of a backpropagation NN requires data sets for which we "know the answer." Thus, we must supply data from the system being modeled for which we have measured the input parameters (the independent variables of the model) as well as the output parameters (the dependent variables of the model). Next, we use a three-layer architecture as shown in Figure 11.3 including an input layer, a hidden layer, and an output layer. Associated with these three layers are two weight matrices that contain the values of the weights associated with the synapses connecting the neurons. Finally, the algorithm includes a learning rate term, α, indicating how much of the weight change to effect on each pass. This is typically a number between 0 and 1. There is a momentum term, ϕ, indicating how much a previous weight change should influence the current weight change. There is also a term indicating within what tolerance we can accept an output as "good."

Algorithm 11.1 - The following algorithm can be used to train an NN.

1. Encoding

Randomly assign values to the network weights. Assign random values between -1 and +1 to the weights between the input and hidden layers, and the weights between the hidden and output layers.

2. Training

A. Forward Pass

i. Compute the hidden layer neuron activations:

$\mathbf{h} = F(\mathbf{i}W_1)$

h = an array (vector) containing the activation signals associated with the hidden layer neurons

I = an array containing the activation signals associated with the input layer neurons

W_1 = a two-dimensional array (matrix) containing the weight values for the synapses between the input and hidden layer neurons

ii. Compute the output layer neuron activations:

$\mathbf{o} = F(\mathbf{h}W_2)$

o = an array containing the activation signals associated with the output layer neurons

h = an array containing the activation signals associated with the hidden layer neurons

W_2 = a two-dimensional array containing the weight values for the synapses between the hidden and output layer neurons

B. Backward Pass

i. Compute the error associated with the neural network:

$\mathbf{d} = \mathbf{o}\,(1 - \mathbf{o})\,(\mathbf{o} - \mathbf{t})$

d = an array of error values associated with each output neuron

o = an array containing the activation signals associated with the output layer neurons

t = an array of values that represent the "correct answers;" these are target values the neural network is striving to replicate

ii. Compute the hidden layer error:
$\mathbf{e} = \mathbf{h}\,(1 - \mathbf{h})\,W_2\mathbf{d}$

e =	an array of error values associated with each hidden layer neuron
h =	an array containing the activation signals associated with the hidden layer neurons
d =	an array of error values associated with each output neuron
W_2 =	a two-dimensional array containing the weight values for the synapses between the hidden and output layer neurons

iii. Adjust the hidden-to-output layer weights (W_2):
$W_2 = W_2 + \Delta W_2$

ΔW_2 = a two-dimensional array containing changes to be made to the weight matrix. These changes are computed according to the formula:

$\Delta W_{2(t)} = \alpha\mathbf{hd} + \phi\,\Delta W_{2(t-1)}$

α =	learning rate which affects the speed of convergence (but, as the convergence rate is increased, so too is the chance of the network converging to a local, sub-optimal solution)
ϕ =	the momentum term. This term is used to allow the changes made in the previous step to affect the changes in the current step. This is done as a way of discouraging convergence to a sub-optimal solution.

iv. Adjust the weights for the first layer of synaptic connections:
$W_1 = W_1 + W_{1(t)}$
where
$W_{1(t)} = \alpha\mathbf{ie} + \phi\,\Delta W_{1(t-1)}$

Repeat the forward and backward passes on all pattern pairs until the output layer error (vector **d**) is within the specified tolerance for each data set and for each neuron.

Once the NN has been sufficiently trained, it can be used to describe the relationship between input data and output values: it can be used to model the system. To do this, a forward pass can be made through a stagnant network (one in which the backward pass is no longer active).

Neural networks can learn any arbitrarily complex nonlinear mapping. This is due to the introduction of the middle or hidden layer. However, there are potential drawbacks to NNs. First, one of the most cited disadvantages of NNs is that they can involve extremely long training times. In modeling situations for which the relationships between the parameters are subtle or elusive, it is not unreasonable to expect an NN to require weeks of training on computer workstations. However, when the alternative is no computer model to drive your adaptive control system, this training period is not an insurmountable obstacle. Second, NNs can "memorize" the training data but fail to extend generalizations to situations for which they have not been specifically trained. Often, this occurs when the NN has too many hidden nodes; a delicate problem since too few hidden nodes and the NN will not train while too many nodes causes the NN to memorize. Third, NNs do not give any indication as to why they do what they do. In other words, when an NN is used as a computer model, it receives input data and produces an output value. There is no easy way of understanding the complex relationships between the input values and output values that are contained within the NNs weight matrices. Thus, NNs are the quintessential "black box" device.

Neural networks are powerful tools for modeling data. In principle, they are easy to use: simply show the NN some training sets (data collected from the system being modeled) and watch it "learn to model the data." However, in practice, the act of training an NN to model a physical system is more of an art than a science. There are several aspects of the NN's architecture (most notably the number of middle nodes) and training (the learning rate, α, and the momentum parameter, θ) that affect the NN's ability to model a system that must be set by a user pretty much by trial and error.

Thus, NNs probably should not be thought of as "the be-all and end-all" of computer modeling as they once were. Neither, however, should they be thought of as more trouble than they are worth. NNs are powerful modeling tools that can be used to model a variety of systems. Their effective use in a particular situation requires some knowledge both about the system being modeled and the NNs themselves. In the remainder of this chapter we will show three specific instances in which an NN has been used to effectively model physical systems.

11.3 NEURAL NETWORK MODEL OF GRINDING

The grinding process is common in the mineral processing industry, and is characterized by several performance measures, all of which are important in various circumstances. Generally, there are four measures that are especially important indicators of the efficiency of a particular grinding process: (1) *fineness* of the ground product, (2) *energy* costs associated with the process, (3) a parameter associated with the *viscosity* of the slurry, and (4) a second parameter associated with the *viscosity* of the slurry containing the ground product. These performance measures all are functions of four basic input parameters: (1) the percent solids by weight of the feed to the grinding circuit, x_S, (2) the maximum ball size, x_B, (3) the mill speed, x_M, and (4) the dispersant addition, x_D. Four individual neural networks have been trained to predict each of the performance measures (output variables) given the values of the four input parameters. The neural networks used to model grinding were classic three layer, backpropagation neural networks with four input nodes, eleven hidden nodes, and a single output node. In the training of each network, a learning rate of 0.2 was used while the momentum term was set at 0.05. Figure 11.4 demonstrates the effectiveness of the neural network model used to predict fineness. In this figure, the model predicted fineness is plotted against a measured value. Realize that if the model were exact, all of the points in this plot would lie on a 45° line. Also, it is important to note that Figure 11.4 consists of data the neural network model had not seen before; it is not the training data.

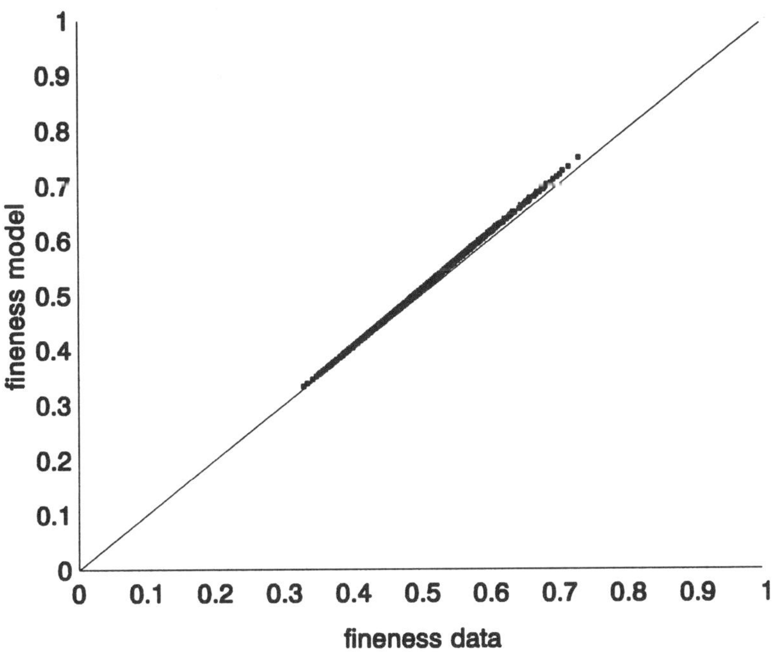

Figure 11.4 - The neural network model predicts fineness quite accurately, as is borne out by a correlation coefficient of 0.963.

The NN is a valuable tool for modeling grinding because the physical phenomena occurring in grinding are not easily modeled with first principle models. The neural network model was trained over a two-week period using 400 training data sets and a 486 personal computer. The two-week development period included the time needed to consider various NN architectures and system parameters (α and ϕ).

11.4 NEURAL NETWORK MODEL OF HYDROCYCLONE

Hydrocyclones are a standard method of classifying slurries in the mineral processing industry. Hydrocyclones are continuously operating separation devices that utilize centrifugal forces to accelerate the separation of particles. They have achieved such popularity because of their simplicity, their durability, and their relatively low cost. Hydrocyclones are now used increasingly in closed-circuit grinding, de-sliming circuits, de-gritting procedures, and thickening operations (Willis 1979).

A typical hydrocyclone is shown in Figure 11.5. The mechanics of this device were originally described in Chapter 10; they are repeated here for convenience. The lower portion of a hydrocyclone is a conical vessel with an opening at the *apex* or bottom to allow for the removal of the coarse or heavier particles. The conical section is joined to a cylindrical section, the top of which is closed with the exception of an overflow pipe known as a *vortex finder*. The vortex finder prevents the tangentially fed mineral from going directly into the overflow, while allowing the fine particles a means of exiting the hydrocyclone. In this cylindrical section the actual separation occurs, due to the existence of a complex velocity distribution that carries the coarse particles to the apex and the fine particles out the top.

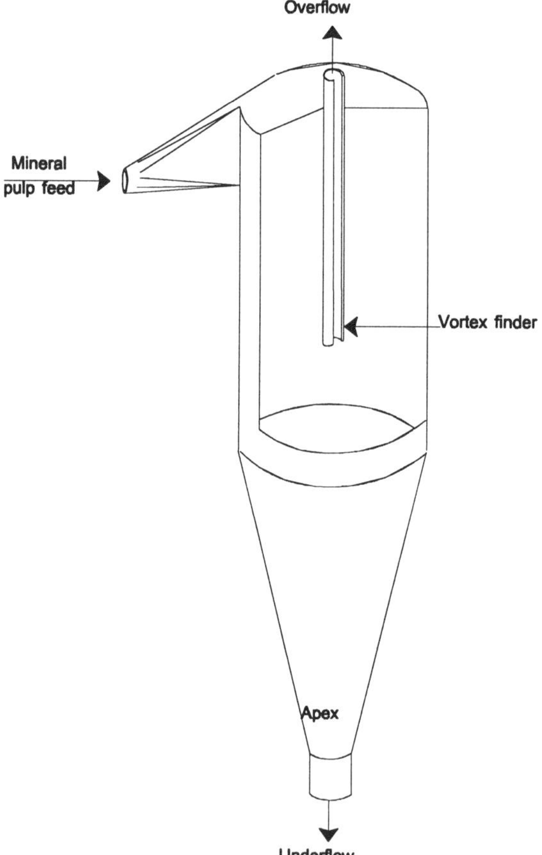

Figure 11.5 - Hydrocyclones are extremely popular separators because of their simplicity, their durability, and their relatively low cost.

Hydrocyclones have traditionally been modeled using empirical relationships. Plitt (1976) developed a model to predict the d_{50} or *split size* that is still used extensively today. The split size is that size particle (given by diameter of the particle) that has an equal chance of exiting the hydrocyclone either through the underflow or the overflow, and is often used to quantify the size separation. Unfortunately, this particular model does not perform well across a spectrum of hydrocyclone sizes. In Chapter 10, we saw how a genetic algorithm could be used to tune such a computer model, but that exercise was dependent on the basic modeling equation, which may or may not be valid across a spectrum of hydrocyclone cyclones. Therefore, a neural network model of a hydrocyclone has been developed and tested. As with the grinding circuit model, the hydrocyclone model is a three-layer, backpropagation neural network. The neural network has eight input nodes, eleven hidden nodes, and one output node. The inputs to the neural network model are: the diameter of the hydrocyclone, D_c, the diameter of the slurry input, D_i, the diameter of the overflow, D_o, the diameter of the underflow, D_u, the height of the hydrocyclone, h, the volumetric flow rate into the hydrocyclone, Q, the percent solids, ϕ, and the density of the solids, ρ. The output of the neural network model is the d_{50} size. Again, the learning rate and the momentum term were set to 0.2 and 0.05, respectively. Figure 11.6 shows the performance of the neural network model. The model predicted d_{50} is plotted against a measured value. The correlation coefficient for the curve is 0.984.

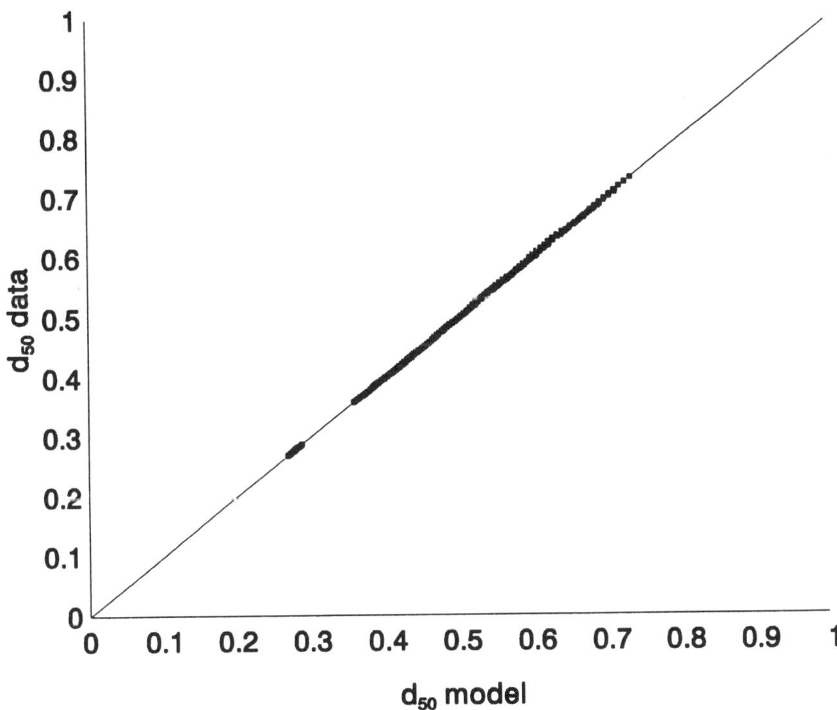

Figure 11.6 - A neural network model of a hydrocyclone predicts the d_{50} size without the necessity of a modeling equation. The correlation coefficient for the curve is 0.984.

11.5 NEURAL NETWORK MODEL OF FLOTATION

Froth flotation is a common method of separating minerals. In froth flotation, minerals are separated when selected particles are rendered hydrophobic (water repellent) with chemical agents (called collectors) and attach to air bubbled through a slurry containing the minerals. These bubble-particle groups then rise to the top of the flotation cell, where they are removed as a froth. Hydrophilic (water attracted) particles are not attracted to the air and settle out separately. In the iron industry, the objective of the flotation process is to remove silica from iron deposits because most commercial uses often require the product to contain less than 6 pct silica. A computer model of flotation could assist in achieving this goal of 6 pct silica through improved system design and process control.

Although several computer models of the flotation process have been developed and tested (Schuhmann, 1942; Tomlinson and Fleming, 1965; Flint and Howarth, 1971), most of the models are effective for only a narrow range of flotation applications. To be effective in the investigation of alternative process control strategies and in the development of improved flotation circuit designs, a computer model must be robust. The robust modeling capabilities of neural networks quite possibly will be adequate to produce a computer model of froth flotation. Such a model could then be used in control systems.

There are a number of system variables that play a role in determining the amount of silica that will appear in the product for iron flotation. However, the purity of the product is particularly sensitive to two parameters: (1) the amount of amine (a collector) used and (2) the feed mass flow rate into the flotation cell. Thus, the fuzzy linguistic model presented in this paper is a two-input (amine flow rate and feed mass flow rate), one-output (percent silica in the product) model of a conventional flotation cell used in the iron industry. The neural network had five hidden nodes, used a learning rate of 0.02, and a momentum value of 0.05. The data used to produce the computer model was acquired from an industrial cooperator; the data was real-world data obtained from an operating plant.

Figure 11.7 shows the performance of the NN model of flotation. The correlation coefficient for this curve is 0.941. Although this model does not match the data as well as the grinding and hydrocyclone models, it still performs adequately for use in the control systems being developed in this book.

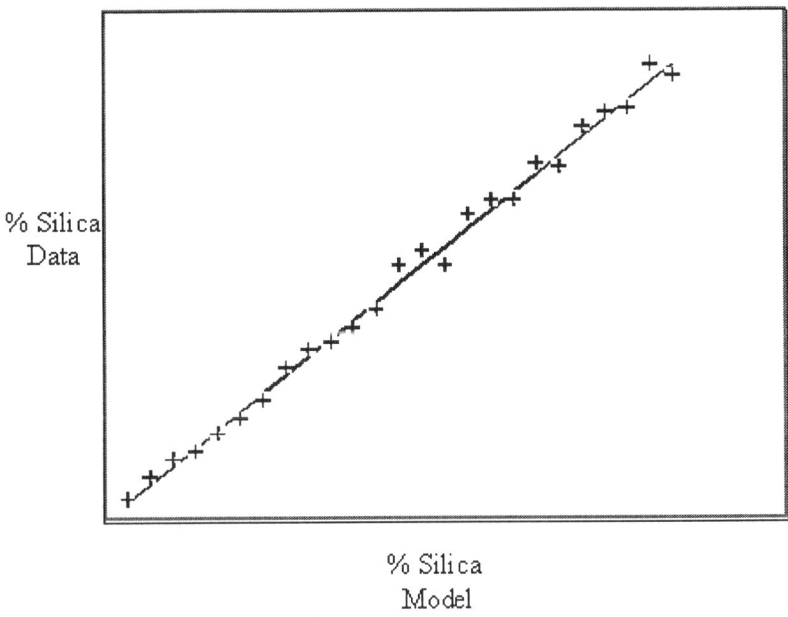

Figure 11.7 - A neural network model of a flotation circuit provides a correlation coefficient of 0.941.

11.6 REVIEW/PREVIEW

Neural networks are computational paradigms based on an analogy with the mammalian brain. They consist of layers of computational nodes connected by weighted synaptic connections. Each computational node receives input signals, sums these signals, passes the signal through a squashing function (a sigmoidal function), and passes its activation function value on to other nodes. Neural networks employ supervised learning in which it is supplied with a set of data cases that contain values of input and output values. These data cases are used to train the neural networks. This approach is different from most modeling techniques that are programmed rather than trained. Backpropagation networks are able to propagate errors calculated using the training cases back through the network and make adjustments to the weights in the network so that the network matches the supplied training cases.

Neural networks are inviting modeling tools because they do not require the user to have any understanding of the system being modeled (although in the authors' experience, it is almost always useful to spend some time becoming informed about the system being modeled so that the process of developing the NN model can be expedited). They simply use data collected from the system, and a bit of user patience in adjusting the neural network parameters, to become effective models. However, there are some aspects of neural networks that are troubling to potential users. The question we hear most is "how do I really know what it is doing?" The answer, we are afraid, is "you don't." This black-box nature of neural networks is simply too much for some users to overcome. Therefore, in the next chapter we will address a modeling technique that is similar to a neural network in that it can be trained, but very different in that the relationships between the input and output parameters is spelled out in terms of a set of rules. In the next chapter we introduce the idea of using fuzzy rule bases to model physical systems.

REFERENCES

Flint, L. R., & Howarth, W. J. (1971). The collision efficiency of small particles with spherical air bubbles. *Chemical Engineering Science*, **26**, 1155-1168.

McClelland, J. L., and Rumelhart, D. E. (1988). *Parallel distributed processing*. Cambridge, MA: The MIT Press.

Pao, Y. (1989). *Adaptive pattern recognition and neural networks*. Reading, MA: Addison-Wesley Publishing Company, Inc.

Schuhmann, R. (1942). Flotation kinetics. I. Methods for steady state study of flotation problems. *Journal of Physical Chemistry*, **46**, 891-902.

Tomlinson, H. S., & Fleming, M. G. (1965). Flotation rate studies. *International Mineral Processing Congress*, **6**, 563-579.

Wasserman, P. D. (1989). *Neural computing: Theory and practice*. New York, NY: Van Nostrand Reinhold.

Willis, B. A. (1979). *Mineral Processing Technology*. Toronto: Pergamon Press.

Chapter 12

Fuzzy Rules for Computer Modeling

12.1 INTRODUCTION

The adaptive fuzzy controllers presented in this book employ computer models of the systems being controlled to provide information about changes in the physical systems — they are *model following control systems*. The adaptive model following control systems we will present in the chapters to come also rely on such computer models to develop and update their control strategies in response to changed conditions. The control strategies are updated using a genetic algorithm to alter the rule set and membership functions employed by a fuzzy control system. Thus, computer modeling is quite important to the work to be presented in the remainder of this book. There is a wide variety of effective techniques available for developing computer models. To date, we have discussed two such methods; in this chapter we present yet another. This chapter discusses the use of fuzzy mathematics for modeling physical systems.

First-principle models are often difficult to develop for industrial systems because of the complex only partially known chemistry and physics associated with many processes. However, there is little doubt that most industries can benefit from the application of computer models. In the separation industry, for instance, unit processes including froth flotation, column flotation, grinding, and leaching are currently operated at less than optimum performance, and the development of accurate computer models represents a major step toward rectifying this situation (Kelly and Spottiswood, 1982; Karr, Yeager, and Stanley, 1994). Accurate computer models can be used to improve economic efficiency in both adaptive control and in equipment design. Because of the potential for economic improvements in the separation industry, as well as in other industries, researchers have begun to focus on the development of modeling techniques (Karr, 1993a; Karr 1993b).

As we saw in Chapter 11, neural networks are quite promising tools for model development in the separations industry. In fact, at first consideration, they seem like a panacea for modeling. Neural networks are not "programmed," they are "trained" (Wasserman, 1989). Neural networks are presented with data collected from the system that exemplifies input/output relationships, and they use this data to become efficient predictors of the output response of the system for a given input. Although this approach is inviting for a variety of reasons, including the fact that the model developer does not have to understand the chemistry and physics of the system being modeled, there is a drawback. A neural network is truly a "black box" computer model. To use the model, the neural network is presented with input and it provides a predicted response yet it can give no indication as to why it prescribes a particular output; one cannot discern the relationship between the input and output parameters. This may be unacceptable.

This chapter provides the reader with an overview of fuzzy models. First, similarities between fuzzy control systems and fuzzy models are discussed. Then, a physical systems from the separations industry (a grinding circuit) is modeled using fuzzy algorithms.

12.2 MODELING VS. CONTROL

In the first part of this book we took a rather extensive look at the construction of control systems using fuzzy algorithms. Now, we turn our attention to using fuzzy mathematics to develop computer models. Fortunately, these two tasks are basically the same. In the development of process control systems, the objective is to take a control action that will efficiently bring a system to a control state. This action is based on specific conditions that exist in the system being controlled. In modeling problems, the objective is to predict the state of a set of dependent variables based on the state of a set of independent variables. Thus, in both instances, the objective is to develop a fuzzy module that describes an optimum surface. This idea is shown in Figure 12.1. In the control application, an efficient control action is a function of the current state of the system. In the modeling application, the dependent variables are a function of the independent variables.

Computer models based on fuzzy mathematics will have the dependent variables in the system being modeled described in linguistic terms as some function of the independent variables. As with fuzzy control systems, the rules incorporated in a fuzzy model should make sense to a human. For instance, consider again the cart-pole system we have worked with in previous chapters. A computer model of the cart-pole system would predict the dependent variables, in this case the future state of the cart-pole system (values of x^{t+1}, x^{t+1}, θ^{t+1}, and θ^{t+1}), based on the independent variables, in this case the current state of the system (values of x^t, x^t, θ^t, and θ^t) and any control actions taken on the system (values of F). Thus, a potential rule for modeling the cart-pole system is:

> IF {**cart position** is *centered*, **cart velocity** is *near zero*, **pole angle** is *vertical*, **pole angular velocity** is *near zero*, and **force** is *positive small*}

THEN {**next cart position** will be *slightly positive*, **next cart velocity** will be *positive small*, **next pole angle** will be *small negative*, and **next pole angular velocity** will be *negative small*}.

Certainly, this rule cannot actually be applied in a meaningful way until the linguistic terms shown in italics are provided with exact meaning using fuzzy membership functions. Thus, both fuzzy controllers and fuzzy models depict a relationship between "input" and "output" variables using linguistic rules and membership functions that define the linguistic terms appearing in the rules.

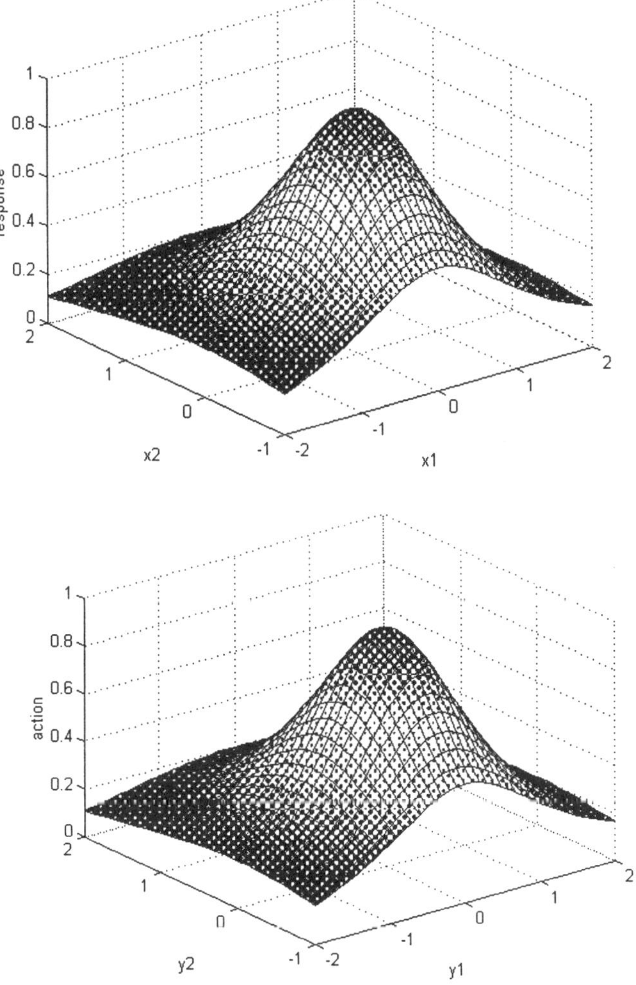

Figure 12.1 - The tasks of developing both control systems and computer models are to define optimum surfaces. In the case of a controller, the correct control action is a function of the state of the system. In the case of a modeling system, the correct future state depends on the current state and any control actions taken on the system.

Once it is understood that computer models define a particular surface just as controllers do, then the steps necessary to develop a fuzzy computer model become apparent. In fact, the steps used for developing a fuzzy controller apply with only slight modification. The following steps can be used to develop a fuzzy model of a physical system:

1) Determine the state variables and the forcing functions (those things that effect changes in the environment to be modeled) to be considered. These are the variables that represent the current state of the system and the forcings that are applied to the system to elicit a change in state.

2) Determine the new state variables to be computed. These are the variables that the model is predicting.

3) Describe the fuzzy sets for the current state variables, the forcings, and the new state variables to be predicted.

4) Establish fuzzy production rules that cover all of the possible conditions that exist in the problem environment.

5) Define the fuzzy membership functions for all descriptive terms to be used in the fuzzy rules.

6) Apply the fuzzy production rules by taking a weighted average of the actions prescribed by all of the rules. The result is a single crisp numeric value for each of the state variables being predicted.

This procedure is almost identical to the procedure set forth in Chapter 2 for developing a fuzzy control system. And, in the next sections, we hope to show that these steps can be used to develop efficient computer models.

12.3 FUZZY MODEL OF A GRINDING CIRCUIT

In Chapter 11 we developed a neural network model of a grinding circuit. At that time, we pointed out some of the highlights of grinding, and some of the rationale for modeling grinding. Since we have already covered these points in some detail, we choose simply to provide a review at this point.

Grinding is a necessary component in the separation of a number of minerals, and improvements in grinding could provide substantial cost savings for the minerals separation industry. Because of the complex physics associated with grinding, it is a particularly difficult unit process to model. Thus, researchers have begun to turn their attention away from the development of first-principle models and toward alternative mechanisms of modeling.

The problem of modeling grinding using fuzzy mathematics is identical to the problem definition set forth in the neural network modeling application. Specifically, the chore is to predict four performance measures: (1) the *fineness* of the ground product, (2) *energy* costs associated with the process, (3) *distribution modulus* that is the packing density at minimal viscosity, and (4) the apparent *viscosity* of the ground product. Each of these four indicators are important in varying industrial situations. Although there are a number of parameters that can alter the performance of a grinding circuit, there are four variables that are especially important: (1) x_s, the percent solids by weight, (2) x_B, the maximum ball size, (3) x_M, mill speed, and (4) x_D, dispersant addition. Thus, the intent of the current effort is to develop a linguistic model in which x_s, x_B, x_M, and x_D are used to predict the fineness of the product, the energy consumed in a grinding circuit, the distribution modulus, and the apparent viscosity of the ground product. Further, only data describing values of these eight variables is to be used in the development of the model. Figure 12.2 shows a schematic of the linguistic computer model.

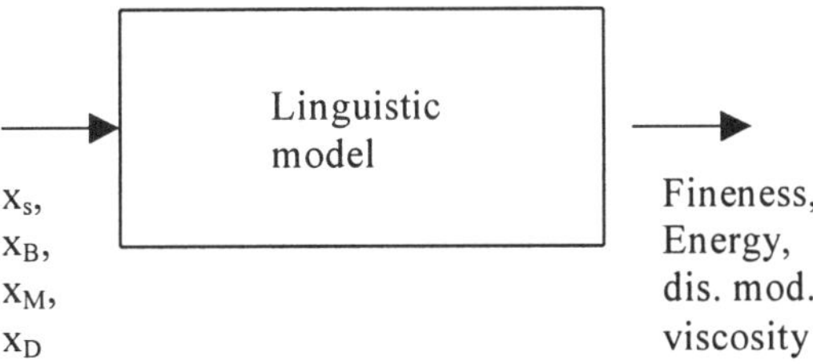

Figure 12.2 - The linguistic model receives as inputs: (1) x_s, the percent solids by weight, (2) x_B, the maximum ball size, (3) x_M, mill speed, and (4) x_D, dispersant addition. It uses values of these parameters to estimate values of: (1) the *fineness* of the ground product, (2) *energy* costs associated with the process, (3) *distribution modulus* that is the packing density at minimal viscosity, and (4) the apparent *viscosity* of the ground product.

Fuzzy linguistic models describe a relationship between input and output variables in linguistic terms. Like conventional rule-based systems (commonly called expert systems), fuzzy linguistic models use a set of production rules that are of the form

IF {*condition*} THEN {*action*}.

The left-hand side of the rules consists of combinations of the input variables (the variables x_s, x_B, x_M, and x_D). The right-hand side of the rules consists of combinations of the output variables (the variables f, e, dm, and v). Unlike conventional expert systems, fuzzy linguistic models discussed here employ rules that utilize fuzzy terms like those appearing in human rules-of-thumb, and like those appearing in fuzzy controllers. For example, a reasonable rule for a fuzzy linguistic model used for a grinding circuit might be:

IF {percent solids is "*small*" AND ball size is "*very smal*"" AND mill speed is "*fast*" AND dispersant addition is "*medium*"}
THEN {fineness is *57.87* AND energy is *16.35* AND distribution modulus is *0.3789* AND viscosity is *0.3227*}.

This rule describes a relationship between the input and output parameters using linguistic terms like "*small*", "*medium*", "*large*", and "*fast*" to represent the input parameters and numeric values to represent the output values.

The linguistic terms are conceptually easy for humans to comprehend. But, these terms are actually subjective linguistic terms — they can mean slightly different things to different human "experts." Such subjective concepts are not readily manipulated by computers, and therefore must be defined using fuzzy membership functions. As with the fuzzy control systems described in Part 1 of this book, the membership functions and the rules must be chosen carefully. The genetic algorithm has been used to expedite the process of rule selection and membership function tuning. The coding and fitness function definitions are straightforward and virtually identical to those set forth in previous chapters.

A self-generating fuzzy linguistic modeling system has been developed that employs a genetic algorithm to select membership functions and to define rule sets. The system is capable of predicting the performance of nonlinear processes and mechanisms by considering only a data set that contains the input/output relationships. Accurate fuzzy linguistic models of both a grinding circuit and a flotation circuit have been generated that provide some insights into the relationships between the various parameters.

Figures 12.3 and 12.4 demonstrate the results of using the fuzzy computer model to predict the fineness associated with data for a training set; a set of data used explicitly by the genetic algorithm to train the model. Figure 12.3 compares the value of fineness predicted by the computer model with the value of fineness of the actual grinding circuit. Note that a "perfect" computer model would elicit points along the 45° line. This "45°" figure provides insight into the accuracy of the computer model. However, it is occasionally difficult to judge the quality of the computer model by considering this plot alone. Figure 12.4 is a histogram that provides an alternative viewpoint on the accuracy of the computer model. This figure gives an indication of the number of points within a certain percent error. Here, percent error (P) is defined according to the following formula:

$$P = \frac{| y_j^{model} - y_j^{data} |}{y_j^{data}} * 100\% \qquad (12.1)$$

where y_j again denotes one of the four output variables (in this instance it is the fineness). Finally, a correlation coefficient has been calculated to further indicate the effectiveness of the computer model. The correlation coefficient for fineness associated with the training set is 0.9599.

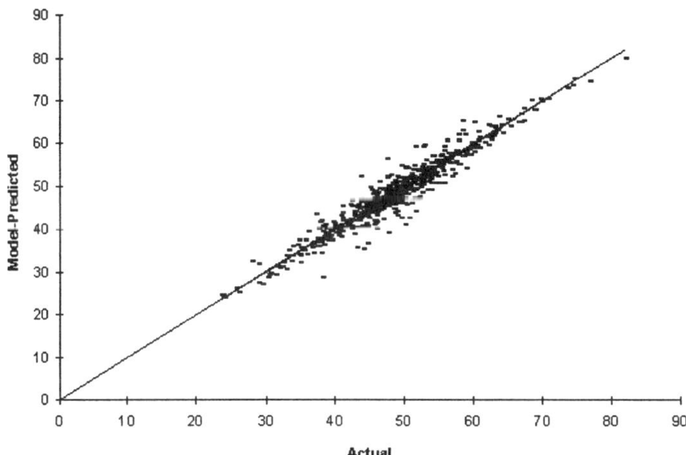

Figure 12.3 - The fuzzy model is trained by a genetic algorithm to accurately estimate fineness on the training data provided.

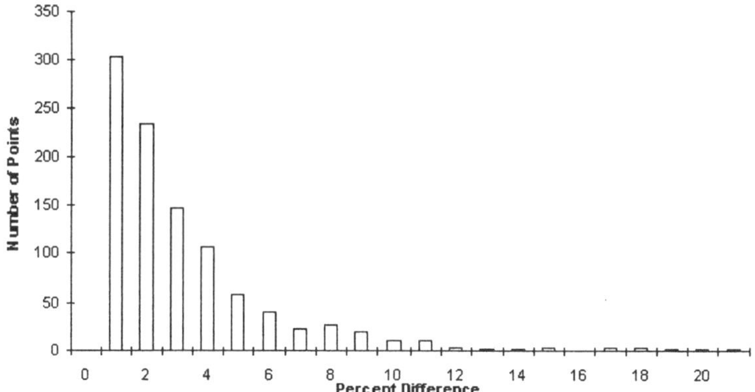

Figure 12.4 - The histogram provides an aid in gauging the performance of the fuzzy model in predicting the data set forth in the training set.

The fuzzy computer model accurately estimates the fineness associated with the training data. However, the real test of the model's effectiveness lies with its ability to compute the value of fineness for a test set containing data the model has never been exposed to. Thus, Figures 12.5 and 12.6 are a "5°" and "histogram" for a test set of data. This test set is composed of data from a grinding circuit and differs from the training set only in that the fuzzy model was not trained on the same data. As with the training set, the fuzzy model is able to accurately compute fineness. A correlation coefficient of 0.9811 was obtained.

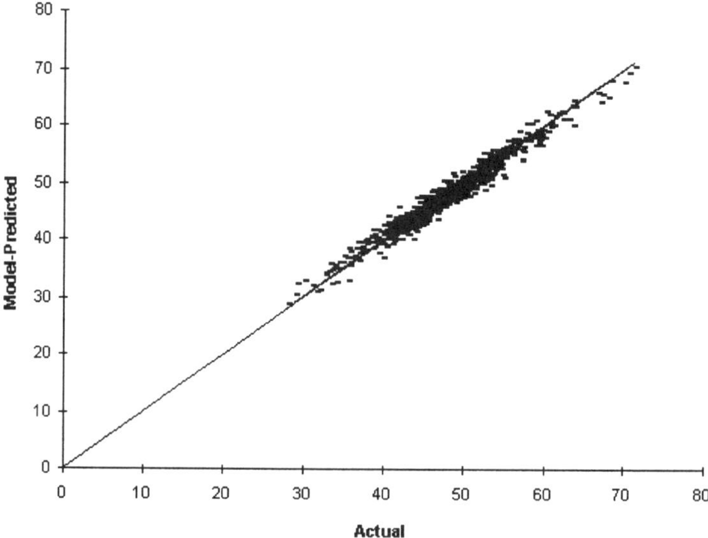

Figure 12.5 - The real test of the fuzzy model comes in its ability to predict values of fineness for which it has not been specifically trained. The above 45° plot represents the performance of the model on test data, not training data.

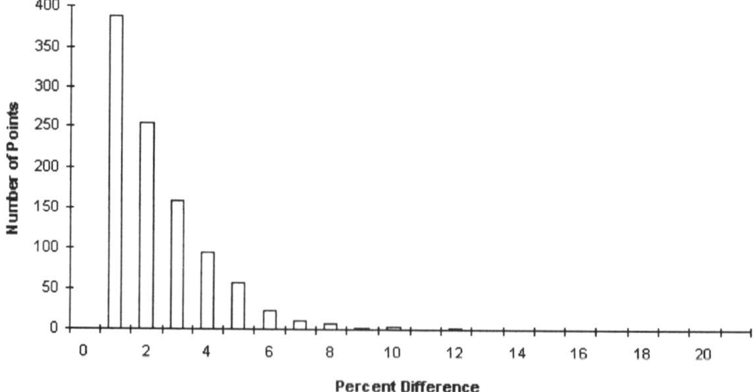

Figure 12.6 - The above histogram shows the performance of the fuzzy model on training data.

The results of the model's ability to estimate fineness as characterized by Figures 12.3 through 12.6 are indicative of its performance in predicting the other three measures of grinding performance. Therefore, for the other three performance measures, (1) *energy* costs associated with the process, (2) *distribution modulus* that is the packing density at minimal viscosity, and (3) the apparent *viscosity* of the ground product, the training cases are not considered. Figures 12.7 through 12.12 show the performance of the model on only the test cases. Additionally, Table 12.1 shows appropriate correlation coefficients.

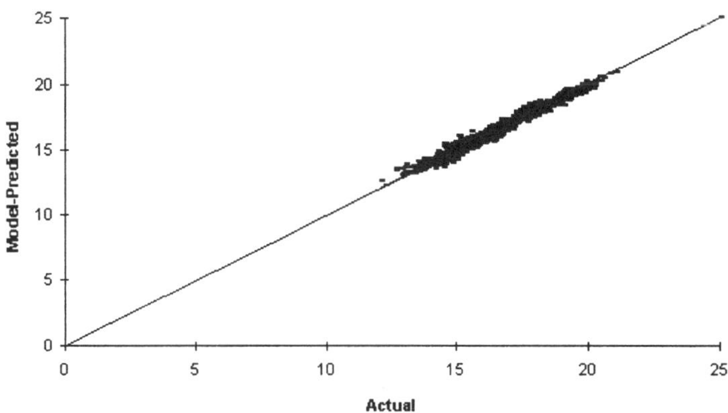

Figure 12.7 - The fuzzy model accurately estimates the energy of the grinding circuit.

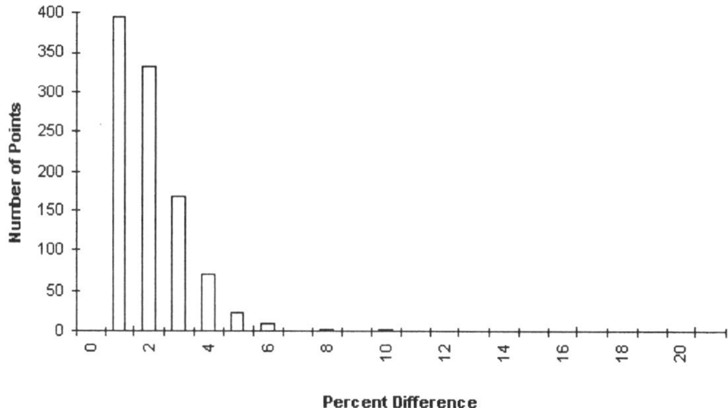

Figure 12.8 - The histogram for energy estimation of test data gives insights into the representative size and distribution of the model error.

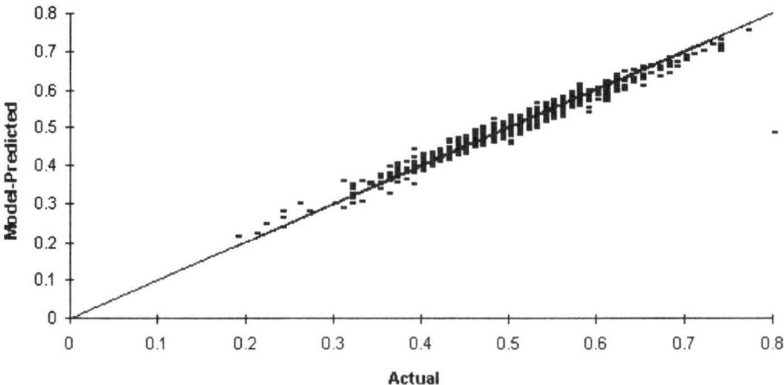

Figure 12.9 - The fuzzy model accurately estimates distribution modulus for test data.

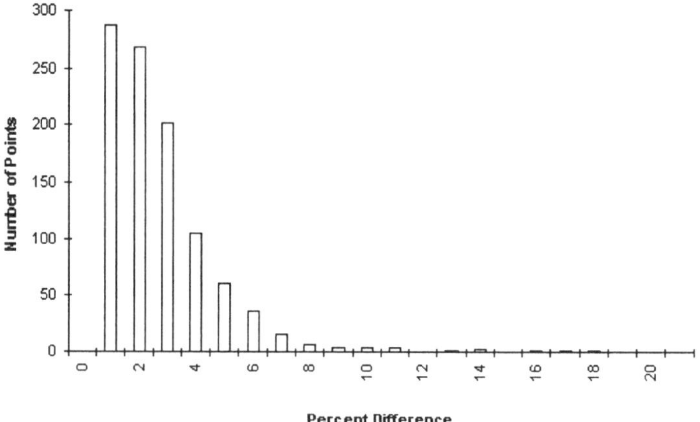

Figure 12.10 - The above histogram reinforces the idea that the model is accurate.

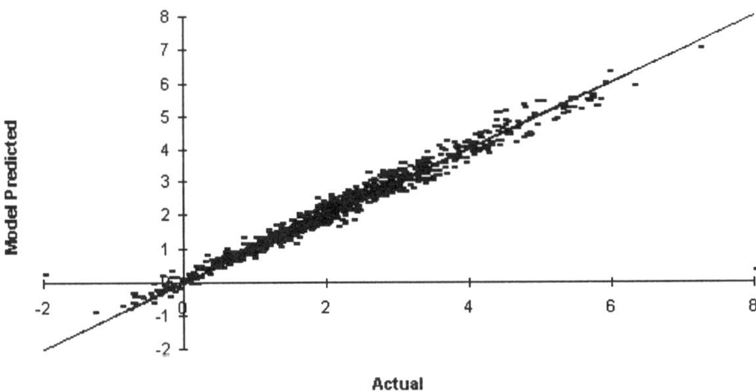

Figure 12.11 - As with the other performance measures, the fuzzy model accurately estimates the viscosity of the ground product for test data.

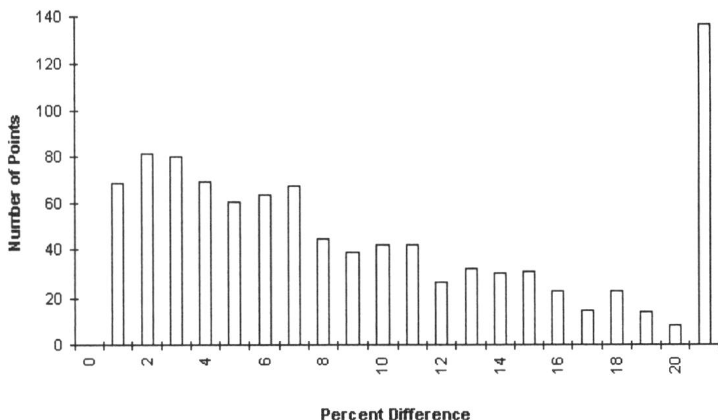

Figure 12.12 - The above histogram for viscosity shows some rather large errors. However, it should be noted that these values are based on a percent difference, and the relative values of viscosity are quite small. The correlation coefficient shown in Table 12.1 alleviates the concern raised in the above figure.

Table 12.1 - The correlation coefficients for the fuzzy model, for both training and test data, show that the computer model effectively predicts the response of a grinding circuit.

Output variable	Training	Test
F	0.9599	0.9811
E	0.9541	0.9852
Dm	0.9572	0.9730
v	0.9730	0.9519

12.4 REVIEW/PREVIEW

In this chapter we have presented the idea of using fuzzy mathematics for developing computer models. Developing a fuzzy model is closely related to developing a fuzzy controller; they both involve the construction of a surface in hyperspace. Thus, the step-by-step procedure for developing a fuzzy controller presented in Chapter 2 was modified to produce a step-by-step procedure for developing a fuzzy model. This procedure was used to develop computer models of two industrial systems: a grinding circuit and a flotation system.

As with fuzzy controllers, one of the delicate issues associated with fuzzy model development is the selection of rules and the tuning of membership functions. If we have done a reasonable job of presenting the material in this book to the current point, the reader should need no encouragement to select a genetic algorithm for this task. And, a genetic algorithm was used in this chapter to tune both the grinding and flotation models.

Fuzzy mathematics offers another approach to developing computer models. This approach has some nice features. The next section of the book provides detailed descriptions in which computer models have been incorporated into controllers to supply the adaptive capabilities needed to effectively control complex industrial systems.

REFERENCES

Karr, C. L. (1993a). Strategy for adaptive process control for a column flotation unit: *Proceedings of the Fifth Workshop on Neural Networks*, **SPIE 2204**, 95-100.

Karr, C. L. (1993b). Optimization of a computer model of a grinding process using genetic algorithms: In H. El-Shall, B. Moudgil, and R. Weigel, *Beneficiation of Phosphate: Theory and Practice*. Littleton, CO: Society for Mining, Metallurgy, and Exploration, 339-345.

Karr, C. L., Yeager, D., and Stanley, D. A. (1994). Tuning empirical computer models using genetic algorithms: *Proceedings of the Society of Computer Simulation Summer Conference*, 236-241.

Kelly, E. G., and Spottiswood, D. J. (1982). *Introduction to mineral processing*, New York, NY: John Wiley & Sons.

Wasserman, P. D. (1989). *Neural computing*, New York, NY: Van Nostrand Reinhold.

PART IV
APPLICATIONS

Chapter 13

Adaptive Control of a pH System

13.1 INTRODUCTION

In Part I of this book, we introduced a procedure for developing fuzzy systems. In Part II, we discussed the use of genetic algorithms for adjusting and tuning fuzzy systems in real time, to produce adaptive systems. In Part III, we provided mechanisms for developing computer models of systems to be controlled. In this final section of the book, we will tie everything together using specific adaptive control systems chosen from different engineering disciplines. The systems considered include a pH system, an exothermic chemical reaction system, a chaotic system, a mineral processing system, and a helicopter.

In the current chapter we will introduce a very general computer software architecture that can be used for adaptive process control. This architecture is quite flexible and is effectively applied to a number of additional systems in subsequent chapters. The architecture builds on concepts and components introduced previously in the book.

Early adaptive controllers simply enhanced conventional non-adaptive control systems. Details of two such strategies for conventional PID controllers appear in articles by Astrom, et al. (1977) and by Clarke and Gawthrop (1981). Unfortunately, the severe demands placed on control systems by industrial processes with rapidly changing dynamics taxed the adaptive PID controllers to their limit, thereby pointing out the need for control systems that were both more powerful and more flexible. The virtual explosion in the use of expert systems has seen them utilized in a number of application areas, and process control has been no exception. The most straightforward, and perhaps the most effective way to utilize an expert system to produce an adaptive controller is to use a rule-based system to alter the gain constant associated with a PID controller. Control systems that adopt this approach were developed by Krauss (1984) and Krauss and Myron (1984) who used an expert system to alter the gain constant in response to changes in the problem environment.

The combination of modern, high-speed computers and artificial intelligence (AI) software resulted in the development of complex control systems that adopted a new approach to adaptive control. This approach is similar to the one used by humans, and performs more efficiently and with greater flexibility than other previous systems. Until now this book has focused on the application of several such techniques to process control and modeling including expert systems, fuzzy controllers, genetic algorithms, and neural networks. These tools can be used to augment conventional control systems and more significantly they can be used to develop entirely new adaptive control system designs.

Two research groups developed adaptive approaches that did not use the GA. In the first effort, Procyk and Mamdani (1978) used a derivative-based method to alter the rules of an FC. In the second effort, Galluzzo, et al. (1991) used an independent set of rules, called metarules, to alter the rule set of an FC. Both of the above are examples in which the control system receives substantial feedback concerning the changes in the problem environment.

GAs have also been introduced into the design of adaptive control systems. Odetayo and McGregor (1989) used a GA to select rules for a conventional rule-based, expert system. Valenzuela-Rendon (1991) also developed a fuzzy classifier system that used a GA to learn a rule set for a controller. Both of these systems failed to address a key point. That is, often because of inadequate sensors or the lack of sensors, systems must be able to cope with meager feedback.

We have developed an approach to the design of adaptive control systems, based on GAs, that is effective in systems in which minimal feedback concerning the state of the problem environment is available. Since the control systems that result must accommodate this lack of feedback, an overall structure is used that is more suitable to the tasks of recognizing, quantifying, and adapting to changes in the problem environment than non-adaptive control systems of the past have been.

The adaptive control architecture we will present in this and following chapters include three components: (1) a ***control element*** to manipulate the problem environment, (2) an ***analysis element*** to recognize changes in the problem environment, and (3) an ***adaptive element*** to adjust to the changes in the problem environment. Each of these components employs either a GA, an FC, a neural network, or some combination of the three techniques. The control element employs a fuzzy controller. The analysis element relies on the combination of a GA and a computer model (which is often a neural network model). The adaptive element combines a fuzzy controller with a GA. Systems based on these three components have been shown to be effective in a number of problem environments (Karr, 1991).

13.2 PROBLEM ENVIRONMENT

The environment used as a vehicle for introducing the three component architecture is a laboratory pH system that is a small-scale representative of pH systems present in a number of industries (Kelly and Spottiswood, 1982; McAvoy, Hsu, and Lowenthal, 1972). The goal of the control system is to drive the pH of a particular solution to a setpoint. The pH system contains nonlinearities and changing process dynamics, and is an extension of a system studied by Galluzzo and co-workers (1991). The nonlinearities occur because the output of pH sensors is proportional to the logarithm of hydrogen ion concentration. The changing process dynamic has three separate causes. First, there is a mechanism for introducing a buffer to alter the manner in which the pH responds as acid or base is added. (The addition of buffer is roughly analogous to increasing the inertia in a mechanical system.) Second, the concentrations of the acid and base that the controller uses to manipulate the pH of the system can be altered. Third, the setpoint of the system, the desired value to which the system pH is to be driven, can be altered.

A schematic of this pH system is shown in Figure 13.1. A beaker initially contains a given volume of a solution having some known pH. There are five valved input streams into the beaker. Only the valves controlling the two ***input control streams*** can be adjusted by the controller. The concentration of these two control input streams is changed by some "random agent" to be either 0.1 M HCl or 0.05 M HCl and 0.1 M NaOH or 0.05 M NaOH. The valves on the other three input streams are used to manipulate ***external streams*** or flows that are altered by the same random agent that manipulates the concentration of the control input streams. Thus, the problem environment can be manipulated by the control system, or by an external agent. Certainly, the control system must be able to control the problem environment despite the changes made by the external agent. Furthermore, the control system has no knowledge of the changes being made by the external agent. Of the three uncontrolled external streams, one is 0.05 M HCl, one is 0.05 M CH_3COONa, and one is a buffer (a combination of 0.1 M CH_3COOH and 0.1 M CH_3COONa). Additionally, the random agent is capable of changing the desired setpoint to which the system pH is to be driven.

The random agent allows for alterations in the system parameters that dramatically change the way in which the problem environment reacts when adjustments are made to the valves on the control streams. Furthermore, the control element has no knowledge of the changes made in the concentrations of the control input streams by the random agent; it is left up to the analysis element to recognize that the concentrations have changed, and to determine what the new concentrations are.

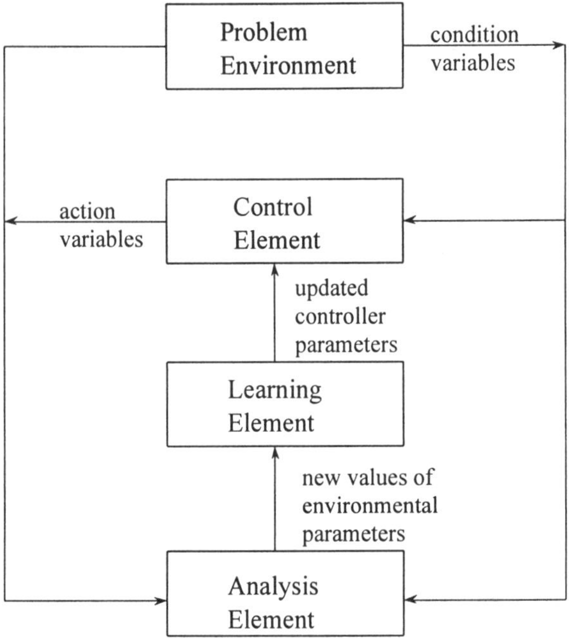

Figure 13.1 – The pH system shown above is a challenging system to control. Aside from the nonlinear nature of pH, the control system is faced with changing conditions including the possible buffering of the system.

The goal of the control problem is to drive the system pH to the desired setpoint in the shortest time possible by adjusting the valves on the two control input streams. Furthermore, the valves on the input streams must be fully closed at the setpoint. As a constraint on the control action, the valves can only be adjusted a limited amount (0.5 mL/s/s, which is 20 pct of the maximum flow rate of 2.5 mL/s), thereby restricting pressure transients in the associated pumping systems.

The pH system was small so that experiments could be performed in limited laboratory space. Titrations were performed in a 1,000-mL beaker using a magnetic bar to stir the solution. Computer-driven peristaltic pumps were used for the five input streams. An industrial pH electrode and transmitter sent signals through an analog-to-digital board to a 33-MHz 386 personal computer that implemented the control system.

13.3 CONTROL ELEMENT

In this section, an FC is used as the control element for the pH system. The details concerning the development of the FC are consistent with the process for

FC design presented earlier in the book, and they are presented in such a way as to make them easily extensible to other problem environments.

As in conventional expert systems, FCs use production rules to arrive at appropriate control operations. The left-hand-side of the rules (the *condition* side) consists of combinations of the controlled variables; the right-hand side of the rules (the *action* side) consists of combinations of the manipulated variables. Unlike conventional expert systems, FCs use fuzzy linguistic terms like those appearing in human rules of thumb. For example, a rule used to manipulate the pH system is:

IF <pH is **VERY ACIDIC** and ΔpH is **SMALL**>
THEN <Q_{BASE} is **LARGE** and Q_{ACID} is **ZERO**>.

This rule says in words that if the solution is very acidic and the pH is not changing rapidly, the flow rate of the base should be made large and the flow rate of the acid should be made zero.

The fact that FCs use fuzzy terms gives rise to another fundamental difference between them and conventional expert systems: a condition variable can be described by more than one fuzzy term. For instance, a system pH of 5 can be described as "very acidic" and "mildly acidic." FCs use fuzzy membership functions to allow the condition variables in several rules to be described, to some degree, by each of the fuzzy terms. Therefore, more than one rule is qualified, or eligible, to enact its action at any given time, i.e., to "fire."

Typically, since more than one rule can have its condition met at any given time, FCs must include a mechanism for determining a single control action. For this purpose a weighted average of the actions prescribed by the appropriate rules is calculated. The emphasis placed on each rule's action is based on the confidence that exists in its condition (the degree to which a concrete value is described by a fuzzy term). The weights used in this averaging technique come directly from the fuzzy membership functions that provide the fuzzy terms with meaning.

Figure 13.2 shows a schematic of the structure of a control element composed of an FC. The control element receives definite values of the condition variables, uses fuzzy membership functions to characterize the definite values with fuzzy terms (it "fuzzifies" the variable values), employs a rule set, and computes *one* definitive action to be taken on the problem environment by calculating a weighted average (it "defuzzifies" the prescribed actions). This process is clarified in the following paragraphs as the membership functions and the rule set used in the pH system are set forth.

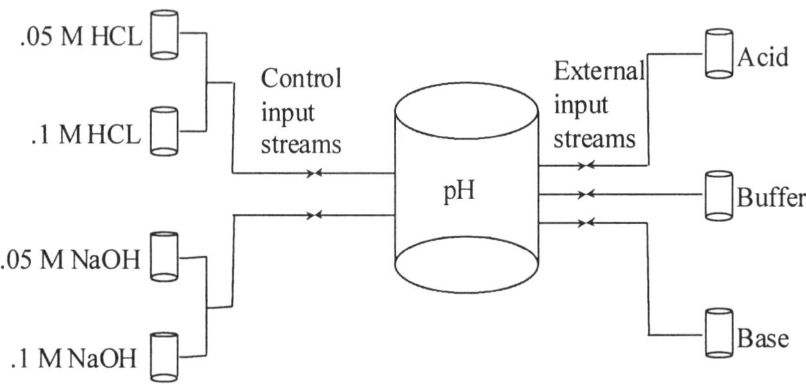

Figure 13.2 – The control element must prescribe effective control actions based on the present condition of the pH environment.

The initial phase of FC development is to determine the appropriate condition and action variables. After a period of experimentation (an inevitable requirement for the development of a quality FC), two condition variables were selected: the current value of pH in the beaker and the absolute value of the current time-rate-of-change of the pH in the tank (ΔpH). Therefore, the two action variables were the flow rates for the acid (Q_{ACID}) and the base (Q_{BASE}), respectively, of the input streams.

Next, fuzzy linguistic variables are selected to represent the condition and action variables. After experimentation, seven terms were selected to describe pH, two terms were selected to describe ΔpH, and four terms each were selected to describe Q_{ACID} and Q_{BASE}. The specific linguistic terms used to describe the pertinent variables in the pH system follow:

pH	Very Acidic (**VA**), Acidic (**A**), Mildly Acidic (**MA**), Neutral (**N**), Mildly Basic (**MB**), Basic (**B**), and Very Basic (**VB**);
ΔpH	Small (**S**) and Large (**L**);
Q_{ACID} and Q_{BASE}	Zero (**Z**), Small (**S**), Medium (**M**), and Large (**L**);

These fuzzy linguistic terms are subjective, but the developer of the pH FC understands what they mean in the context of the physical system to be controlled.

The developer's understanding of the linguistic terms is defined by the membership functions. The initial membership functions used in the FC appear in Figure 13.3. Figure 13.3a depicts the membership functions for pH, Figure 13.3b shows the membership functions for ΔpH, and Figure 13.3c depicts the membership functions for Q_{ACID} and Q_{BASE}. These membership functions are later altered by the adaptive element, via a GA, in response to changes occurring in the pH system. As will be seen, alterations in these functions can dramatically change the performance of the FC.

Although the laboratory pH system is complex and nonlinear, an effective pH FC can be written that contains only 14 rules. The 14 rules are necessary because there are seven fuzzy terms describing the pH and two fuzzy terms describing ΔpH (7*2=14 rules to describe all possible combinations that could exist in the pH system as described by the fuzzy terms represented by the membership functions shown). Each rule is of the form:

IF {pH is A and ΔpH is B} THEN {Q_{ACID} is C and Q_{BASE} is D}.

Thus, each of the 14 rules prescribe two actions. The entire rule set for the pH FC is shown in Figure 13.4. In this figure, the left table shows the actions for Q_{ACID} while the right table shows the actions for Q_{BASE}.

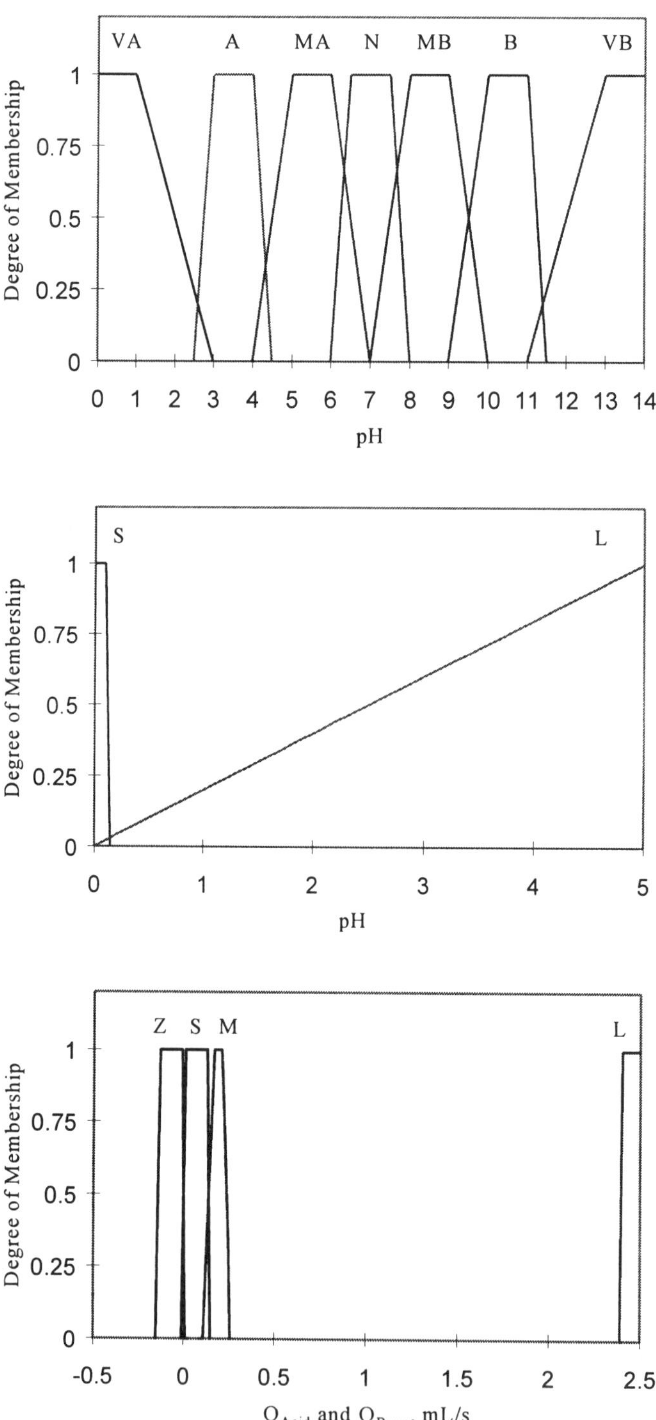

Figure 13.3 – The membership functions shown above represent the developer's rough conception of the linguistic terms used in the control element.

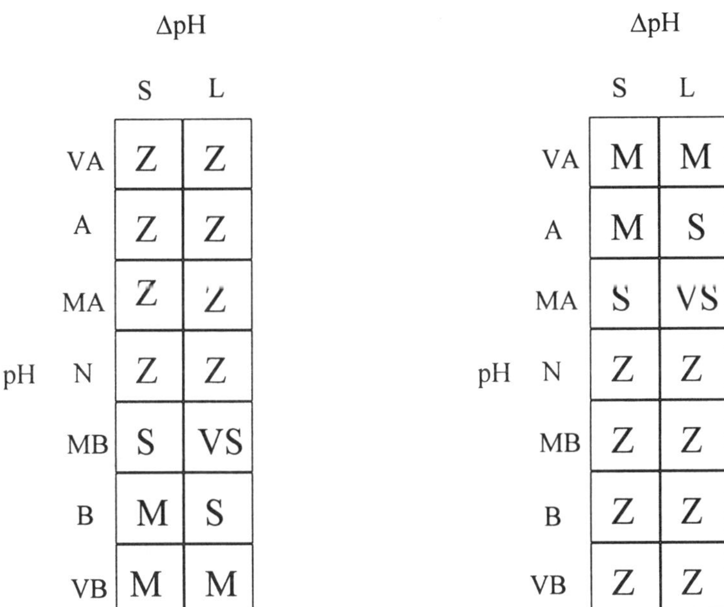

Figure 13.4 – The 14 rules shown above prove to be adequate to control the pH system when all of the environmental changes are disregarded.

Now, the only aspect of the initial control element design that is left is the technique for determining *one* value for the flow rates of the input acid and base streams. The center-of-area (COA) method (Larkin, 1985) discussed in the early chapters of this book is used again. This method provides a convenient way to compute a weighted average of the different control actions prescribed by the rules that are eligible to fire. The COA method results in the selection of a single control action to be taken on the problem environment.

One detail associated with the pH system being considered warrants special mention. There is a limit on the allowable change in the flow rates of the input streams, i.e., the flow rates cannot change by more than 0.5 mL/s/s. But, the membership functions describing the action variables used in the COA method (shown in Figure 13.3) allow for values of Q_{ACID} and Q_{BASE} to range between 0.0 and 2.5 mL/s. The constraint is imposed by computing the value of the flow rates using the COA method. If the computed value is such that it exceeds the constrained flow rate, then the flow rate is changed by the maximum allowable value of 0.5 mL/s.

Figure 13.5 shows the effectiveness of a fuzzy controller in the pH problem. The control element is able to drive the pH to a setpoint of 7 as long as the pH system remains unbuffered. However, when the system is buffered, the control element is not effective because the pH system has undergone substantial changes and the rules in the control element need to be modified to account for such changes.

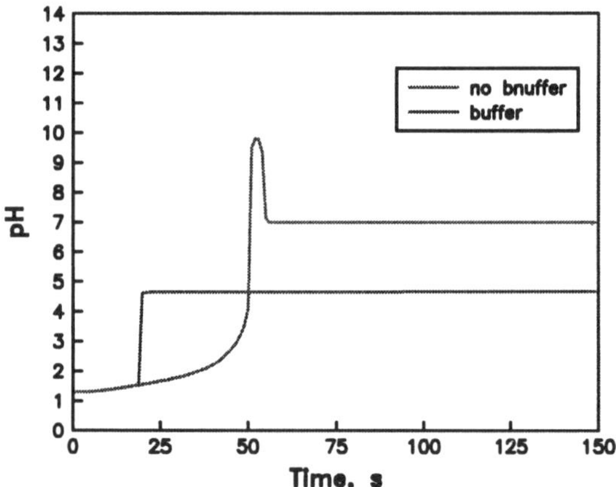

Figure 13.5 – The fuzzy controller is capable of neutralizing the pH in the unbuffered system.

13.5 ANALYSIS ELEMENT

The analysis element must recognize changes in parameters associated with the problem environment that are not taken into account by the rules of the control element. These parameters include: (1) the concentrations of the acid and base of the input control streams, (2) the flow rates of the acid, the base, and the buffer that are altered by an external agent, and (3) the system setpoint. Changes to any of these parameters can dramatically alter the way in which the system pH reacts to additions of acid or base. Recall that the FC used for the control element presented includes none of these parameters in its 14 rules. Therefore, some mechanism for altering the prescribed actions must be included in the control system. But before the control element can be altered, the control system must recognize that the problem environment has been changed, and compute the nature and magnitude of the changes.

Fortunately, the dynamics of the pH system are well understood for buffered reactions, and can be modeled using a single cubic equation (Hand and Blewitt, 1986) that is solved for $[H_3O^+]$ ion concentrations to directly yield the pH of the solution:

$$x^3 + Ax^2 + Bx + C = 0 \tag{13.1}$$

where $x = [H_3O^+]$
$A = k_a + [CH_3COONa] + [NaOH] - [HCl]$,
$B = k_a[NaOH] - k_a[HCl] - k_a[CH_3OOH] - k_w$,
$C = -k_a k_w$,
$k_a = 1.8*10^{-5}$ is the equilibrium constant for CH_3COOH,
$k_w = 1.0*10^{-14}$ is the equilibrium constant for H_2O,

and bracketed terms ([ ]) represent molar concentrations. Further details of the computer model appear in a paper by Karr and Gentry (1992). In physical systems for which the physics are not as clearly understood, alternative modeling techniques such as those discussed in Part III of this book can be employed.

Figure 13.6 shows a schematic of an analysis element. In the approach represented by the schematic, a computer model mirrors the actual problem environment. The response of the physical pH system is compared with the response as predicted by the computer model. When these responses differ by some threshold amount over a substantial period of time, the parameters of the pH system have changed and the model must be updated. Certainly, the threshold and the "substantial" period of time depend on the problem environment. For the pH system considered, when the predicted pH differed from the actual pH of the physical system by 1 unit of measure or more for a period of 5 seconds, the pH system parameters are considered to have changed and new parameters are required. The problem of computing the new system parameters is a curve-fitting problem. The parameters must be recalculated so that the response of the model matches the response of the actual problem environment.

In chapter 10 we saw that GAs can be effective parameter tuners when the basic equation of the model is known. Thus, a simple GA has been used here to tune the parameters. At this point, an analysis element has been forged in which a GA, as described earlier in the book, is used to compute the model parameters necessary to accurately predict the response of the laboratory pH system. A binary coding consistent with that described in Part II of this book was used to form 200-bit strings representing the appropriate model parameters. The first 40 bits of the strings were used to represent the concentration of the acid on the control input stream, the second 40 bits were used to represent the concentration of the base on the control input stream, the third 40 bits were used to represent the flow rate of the acid of the external streams, and the final 80 bits were used to represent the flow rates of the buffer and the base of the external streams, respectively.

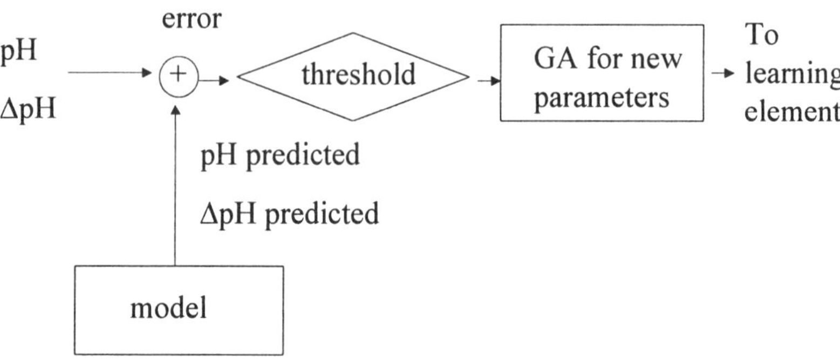

Figure 13.6 – The analysis element tracks the performance of the real-world environment and compares it with a computer simulation. When the two performances differ by a substantial amount, then the real-world system is considered to have changed.

An effective fitness function to select model parameters that mimic the response of the laboratory pH system is:

$$f = \sum_{i=0s}^{i=100s} | pH_{model} - pH_{actual} | . \tag{13.2}$$

With this definition of the fitness function, the problem becomes a minimization problem: the GA must minimize f, which as it has been defined, represents the difference between the response predicted by the model and the response of the laboratory system.

Figure 13.7 illustrates the ability of a GA to locate the parameters needed by the computer model to calculate the response of the physical pH system. This figure includes information concerning the performance of the GA in locating these parameters. The GA was able to locate the correct parameters after only 3000 function evaluations. Locating the correct parameters took approximately 5 minutes on a 386 personal computer. The physical system often mandates that a control action be taken in less than 5 minutes. In this case, the time the GA is allotted to update the model parameters can be restricted. In such situations, the model simply must operate with inaccurate parameters until the analysis element is again employed. However, the magnitude of this problem will be diminished as computers become ever faster. And, the execution speed is adequate for the pH control system.

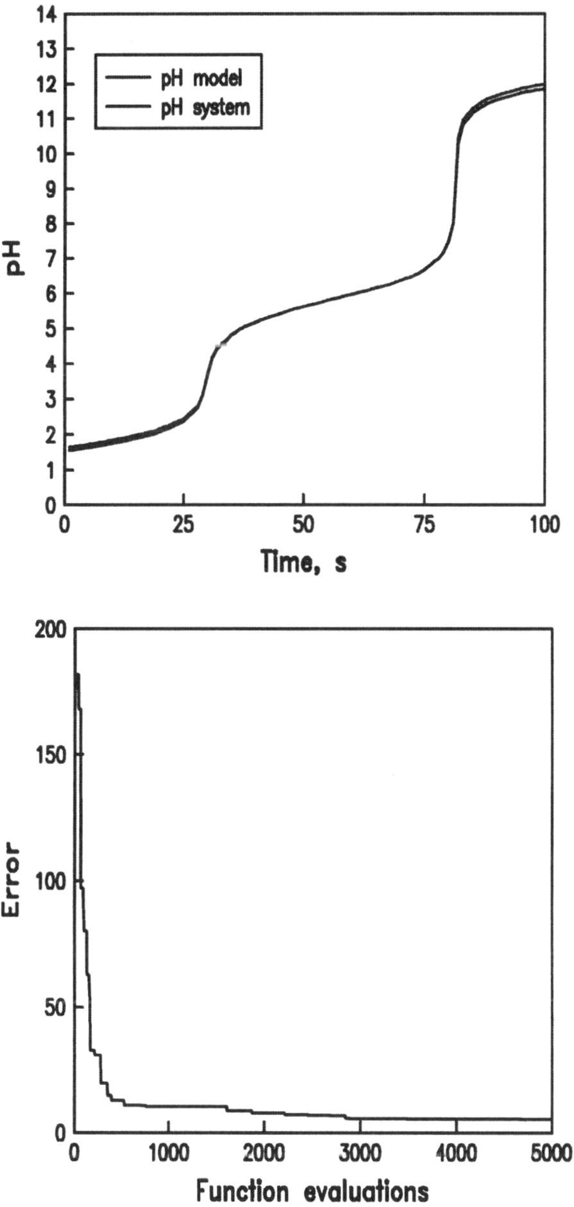

Figure 13.7 – The GA is capable of solving the curve-fitting problem that arises in the operation of the analysis element.

The purpose of the analysis element is to recognize changes in the parameters associated with the problem environment that are not accounted for by the control element, and to compute the new values of these parameters. Once new parameters (and thus the new response characteristics of the problem environment) have been determined, the adaptive controller must alter the control element.

13.6 ADAPTIVE ELEMENT

The adaptive element is responsible for altering the control element in response to changes in the problem environment. Recall that the relevant changes occurring in the pH system include: (1) changes in the concentrations of the acid and base of the control input stream, (2) random additions of acid, base, and buffer from the external streams, and (3) changes in the system setpoint. As set forth in a previous section, none of the parameters associated with the above changes are included in the rule set of the FC that serves as the control element. Therefore, the only way to account for these conditions (outside of completely revamping the system) is to alter the membership functions employed by the FC. The approach we have developed and implemented for using a GA to alter the membership functions associated with an FC has been well documented in previous chapters of this book. Thus, only the details for the pH system implementation are provided here.

A GA used bit strings of length 224 to represent the 32 parameters associated with the search problem. The 32 parameters represent the fuzzy membership functions shown in Figure 13.3 (symmetry considerations allowed for a reduction in the number of parameters that had to be represented). The fitness function must reflect the goal(s) of the control system. In the pH system, the objective is to drive the system pH to a desired setpoint in the shortest time possible, and to keep it there. The fitness function used in this application is:

$$f = \sum_{i=0s}^{i=100s} | setpoint - pH | \tag{13.3}$$

where the summation is performed over a 100-second time period as simulated using the mathematical model of the system has been updated by the analysis element. This simulation is initiated from the current state of the laboratory system, i.e., the current values of pH, Q_{ACID}, and Q_{BASE}.

13.7 RESULTS

Consider first a situation where a buffer is added to the pH system randomly. The adaptive pH controller alters the membership functions it uses to enact its production rules (which do not change) although the process dynamics are altered when the buffer is added. This approach is similar to the subconscious actions of a human controller; a human changes his or her definition of the linguistic terms being used in conjunction with their informal rule-of-thumb approach. Figure 13.8 compares the performance of the adaptive GA-FC with a non-adaptive FC. The adaptive controller is able to achieve the objective much more efficiently than the non-adaptive FC because the adaptive controller is flexible enough to accommodate the changing process dynamics.

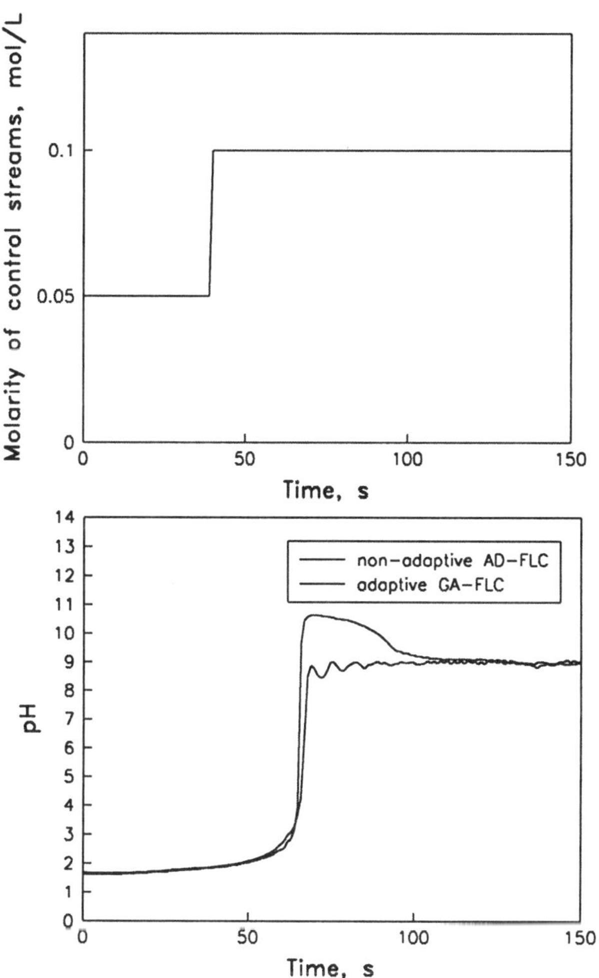

Figure 13.8 – The adaptive controller is able to adjust to changes in the environment and thus out-performs the non-adaptive control system.

Next, consider a situation where the setpoint is changed. The adaptive pH controller must alter its membership functions in response to this change. Realize that declaring a new setpoint is actually changing the objective of the FC. Changing the objective of the controller often requires a modification of the FC rule set. However, the technique of using a GA to alter a set of membership functions is powerful enough to allow the FC to maintain a suitable level of control over the pH system by altering only the meaning of the fuzzy linguistic variables. Figure 13.9 compares the performance of an adaptive GA-FC with a non-adaptive FC. As in the previous example, the adaptive pH controller out-performs the non-adaptive FC.

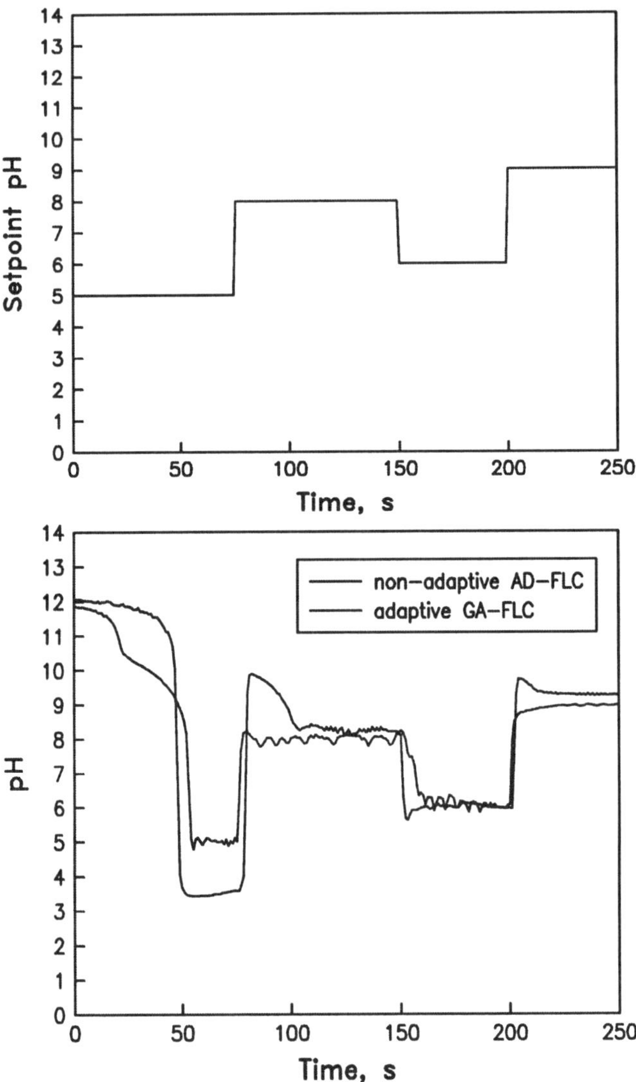

Figure 13.9 – The adaptive controller is able to overcome prescribed changes in the setpoint of the pH system.

Finally, consider a very disruptive change to the pH system, a case where the concentrations of the acid and base the FC is using to manipulate the pH are altered. This is perhaps the most severe change in process dynamics that could be implemented. The response of the system is now completely different: additions of acid or base induce changes in the pH of the system that are far different from the changes in pH that the very same additions of acid or base induced before their concentrations were changed. Figure 13.10 compares the performance of the adaptive pH controller with the performance of a non-adaptive FC. The adaptive GA-FC is able to attain a high degree of control over the pH system despite the dramatic changes in the environment.

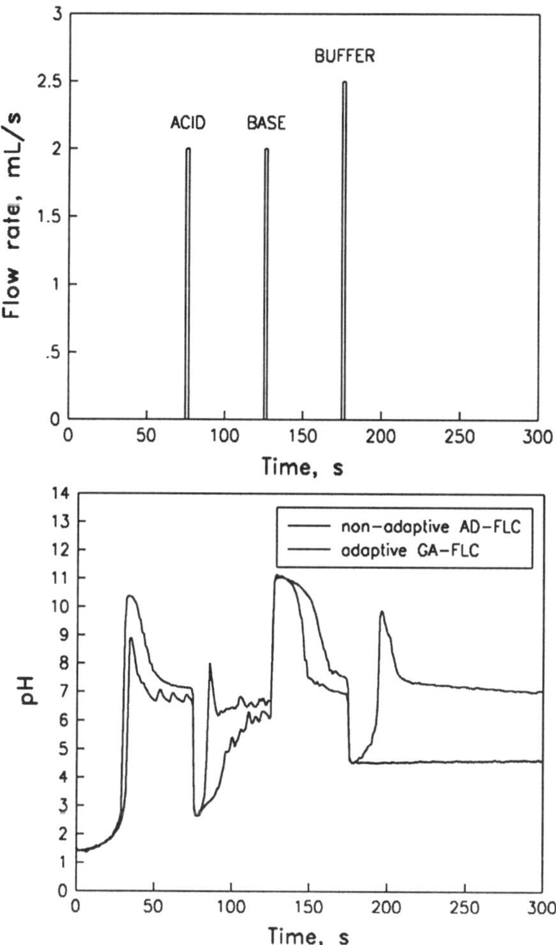

Figure 13.10 – The discrepancies between the adaptive and non-adaptive control system is emphasized when the pH system is buffered.

13.8 SUMMARY AND CONCLUSIONS

We have described an AI-based strategy for adaptive process control. This strategy uses GAs to fashion the three components necessary for a robust, comprehensive adaptive process control system: (1) a control element to manipulate the problem environment, (2) an analysis element to recognize changes in the problem environment, and (3) a learning element to adjust to changes in the problem environment. In this chapter, the strategy has been applied to the development of an adaptive controller for a laboratory acid-base pH system in which the process dynamics change in several different ways. Initially, the overall makeup of an adaptive control system was described. Next, the pH problem environment was introduced. Finally, the basic structure of each of the three individual components was developed, and results were provided demonstrating the merit of using GAs to drive the three components.

The results presented in this chapter demonstrate the power of adaptive control systems based on GAs and FCs. These adaptive control systems are able to recognize when the physical system has changed, to quantify the changes in the physical system, and to maintain a high degree of control over the physical system despite drastic changes in the system characteristics. Based on the results presented, it is concluded that adaptive GA-FCs allow industrial pH systems to be controlled via on-line changes to the membership functions used in the rule base associated with the control system.

Adaptive control systems are becoming vital to the efficient operation of today's industrial plants. If the efficiency of such control systems is to increase, researchers must focus on the synergism of techniques from various fields of study. In this light, the field of AI contains a vast number of untapped resources. Specifically, GAs and FCs demonstrate characteristics that allow for the production of control systems that mimic the approach adopted by humans to the task of process control. And, in the final analysis, humans actually perform the task of adaptive control quite well, as does the adaptive control system presented in this chapter.

In the next two chapters we will apply the software architecture that was applied to a pH system to very different chemical processing systems. In Chapter 15 we will consider the adaptive control of hexamine production in which an exothermic chemical reaction must be effectively manipulated. In Chapter 16 we will apply the adaptive control strategy to a column flotation unit used to separate minerals. In both instances, the adaptive control system is effective.

REFERENCES

Astrom, K. J., Borisson, U., Ljung, L., and Wittenmark, B. (1977). Theory and applications of self-tuning regulators. *Automatica*, **13**, 457-463.

Clarke, D. W., and Gawthrop, P. J. (1981). Implementation and application of microprocessor-based self-tuners. *Automatica*, **17**, 79-85.

Galluzzo, M., Cappellani, V., and Garofalo, U. (1991). Fuzzy control of pH using NAL. *International Journal of Approximate Reasoning*, **5** (6), 505-519.

Hand, C. W., and Blewitt, G. L. (1986). Acid-base chemistry. New York, NY: Macmillan Publishing Company.

Karr, C. L. (1991). Genetic algorithms for fuzzy logic controllers. *AI Expert*, **6** (2), 26-33.

Karr, C. L., and Gentry, E. J. (1992). *Real-time pH control using fuzzy logic and genetic algorithms*. Society of Mining Engineers, AIME preprint 92-49.

Kelly, E. G., and Spottiswood, D. J. (1982). *Introduction to mineral processing*. New York, NY: John Wiley & Sons.

Krauss, T. (1984). Self-tuning control: An expert system approach. *Advances in Instrumentation*, **39**, 1984, 695-703.

Krauss, T., and Myron, T. J. (1984). Self-tuning PID controller uses pattern recognition approach. *Control Engineering*, June, 451-467.

Larkin, L. I. (1985). A fuzzy logic controller for aircraft flight control. Chapter in *Industrial Applications of Fuzzy Control*, M. Sugeno (Ed.). Amsterdam: North-Holland, 87-104.

McAvoy, T. J., Hsu, E., and Lowenthal, S. (1972). Dynamics of pH in controlled stirred tank reactor. *Industrial Engineering and Chemical Process Design Developments*, **11**(1), 68-70.

Odetayo, M. O., and McGregor, D. R. (1989). Genetic algorithm for inducing control rules for a dynamic system. Chapter in *Proceedings of the Third International Conference on Genetic Algorithms*, D. Schaffer (Ed.), Proceedings of a conference held at George Mason University, June 4-7, 1989. Morgan Kaufmann Publishers, 177-182.

Procyk, T. J., and Mamdani, E. H. (1978). A linguistic self-organizing process controller. *Automatica*, **15**, 15-30.

Valenzuela-Rendon, M. (1991). The fuzzy classifier system: A classifier system for continuously varying variables. *Proceedings of the Fourth International Conference on Genetic Algorithms*, R. K. Belew, and L. B. Booker (Eds.), Proceedings of a conference held at San Diego State University, CA, July 13-16, 1991. Morgan Kaufmann Publishers, 346-353.

Chapter 14

Adaptive Control of a Hexamine Production System

14.1 INTRODUCTION

In the previous chapter we saw how a generic software architecture could be used to achieve adaptive control of a pH system. In this chapter we will begin to demonstrate that this architecture is robust by applying it to a different system. As with the pH system, the hexamine system addressed in the current chapter is from the field of chemical processing. However, this exothermic system offers an arena in which multi-objective control problems can be considered.

As has been seen throughout this book, rule-based control and modeling systems incorporate fuzzy linguistic variables into their rule set to model a human's "rule-of-thumb" approach to problem solving. FCs include rules to direct the decision process and membership functions to convert digital values from process sensors into the linguistic values used in the rules. Since both the rule set and the membership functions play an integral role in determining the control action prescribed, an FC can be provided with adaptive capabilities by altering either the rule set or the membership functions, or both. (Recall that in the previous chapter we discussed only the alteration of the membership functions.)

Attaining and maintaining control of exothermic chemical reactions is a problem that demonstrates the need for an adaptive controller. The difficulty of effectively maintaining these exothermic chemical systems at a constant temperature is apparent in a number of diverse industries including mineral and chemical processing (Fogler, 1986; Kelly and Spottiswood, 1982). Adaptive controllers are well suited to the managing nonlinearities and changing process dynamics associated with these chemical systems. The nonlinearities associated with the current chemical system occur in the reaction rate equations and the changing process dynamics are associated with frequent alterations in the concentrations of the input streams. Despite the difficult nature of the problem, FCs have been successfully used for the efficient manipulation of exothermic chemical systems as we saw in the previous chapter and as has been demonstrated in other places (Karr and Gentry, 1993).

The technique for producing adaptive FCs with a genetic algorithm that we present in this book can be employed to alter both membership functions and rule values. In this chapter, the software architecture is applied to a chemical engineering problem. An exothermic chemical system is simulated on a computer. Specifically, a system is considered in which hexamine is produced from ammonia and formaldehyde. This chemical reaction is considered because of its relatively simple kinetics (Kermode and Stevens, 1965), and because the exothermic reaction is typical of the chemical reactions commonly controlled in the chemicals industry.

14.2 PHYSICAL SYSTEM

The problem environment of interest in this chapter is a stirred tank reactor in which ammonia and formaldehyde are reacted to produce hexamine and water. Since the reaction is exothermic, a heat exchanger is included to limit the temperature in the reactor. A schematic of the physical system is shown in Figure 14.1. The objective of the control problem is to maintain a target temperature in the reactor by adjusting the flow rates of both the ammonia into the reactor and the water through the heat exchanger, in response to changes in the flow rate of the formaldehyde. A mathematical model of the physical system is highly desirable to efficiently investigate the control problem. The mathematical model used in this research employs the approach described by Kermode and Stevens (1965). Specifically, the system is modeled with the following set of equations:

Energy

$$q_A \rho C_p T_{Ai} + q_F \rho C_p T_{Fi} - (q_A + q_F) \rho C_p T_{tank} + r(-\Delta H)V - UA\Delta T_m = V\rho C_p \frac{dT}{dt}$$

Mass Balances

$$q_A C_{Ai} - (q_A + q_F) C_A - rV = V \frac{dC_A}{dt}$$

$$q_A C_{Ai} - (q_A + q_F) C_A - rV = V \frac{dC_A}{dt}$$

Heat of reaction

$$-\Delta H = 16610 + 121(T_{tank} - 293.2)$$

Rate of reaction

$$r = kC_A C_F^2$$

where q is the volumetric flow rate (l/sec), C is concentration (moles/l), r is the rate of reaction (moles of ammonia/l s), C_p is the specific heat at constant pressure (gm

K/sec), V is the volume of the reactor (tank) (l), T is the temperature (K), ΔH is the heat of reaction (cal/gm mole), U is the heat transfer coefficient (cal/cm^2 K sec), ΔT_m is the mean temperature of the fluid in the heat exchanger (and is a function of the volumetric flow rate of water through the heat exchanger, q_w), and k is the reaction velocity constant given by:

$$k = 1420 * e^{\frac{-3090}{T_{tank}}}$$

and the subscripts (A and F) indicate ammonia and formaldehyde, whereas the subscript i represents material entering the reactor. The assumptions associated with this model include perfect mixing in the reactor, no heat losses, all physical properties the same as water, and a third-order, irreversible reaction.

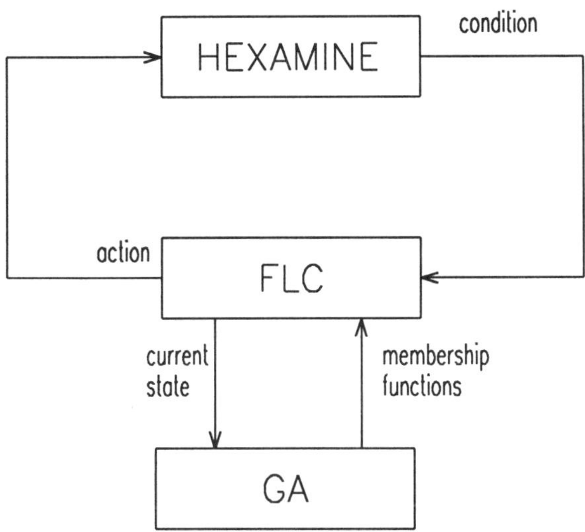

Figure 14.1 - The hexamine system consists of a chemical reactor, inflows of ammonia and formaldehyde, outflows of hexamine and water, and a heat exchanger.

A reactor having a volume of 92.4 liters (l) was simulated. The inflows of ammonia and formaldehyde were allowed to reach maximum values of 1.885 l/sec, while the maximum flow rate of the heat exchanger was 1.2 l/sec. Although the primary objective was to maintain a desired reactor temperature, there was an additional objective. Namely, the production of hexamine was to be maximized. In this research, the desired reactor temperature was 315.0 K. Furthermore, a constraint was placed on the amount the valves controlling the inflow of ammonia can be opened or closed during a given time step. The maximum rate at which the flow of ammonia could be changed was 0.1885 l/s/s. This constraint was enforced to limit transients in the system.

It is a non-trivial task to maintain the temperature in the reactor for various forcing functions (as defined by the rate at which the formaldehyde enters the reactor), much less to ensure the process proceeds efficiently (maximum hexamine production with minimal waste in ammonia and formaldehyde). The current industry standard for controlling a process like the production of hexamine is a proportional-integral-derivative (PID) controller. Although a PID controller can be used to maintain the reactor temperature, it cannot be used to "optimize the process." The GA-driven FC, on the other hand, is capable of both maintaining control of a process and ensuring that it runs at a near-optimum level.

The architecture presented in the previous chapter is used here. The adaptive GA-FC simultaneously utilizes the process control capabilities of fuzzy logic and the search capabilities of GAs. While a FC is manipulating the hexamine system, a GA is searching for improved membership functions (to be used by the FC) to accommodate the changing process dynamics. Once a computer model of the hexamine system has been developed and a crude FC has been written, a GA is employed to produce an efficient, robust, adaptive controller.

14.3 AN ADAPTIVE HEXAMINE GA-FC

At this point, most of the groundwork is laid for the development of an adaptive hexamine GA-FC. The mechanics of a GA were introduced earlier in the book, and the basic coding scheme that will be used has been described and used in the preceding chapter. Before the adaptive capabilities supplied by the analysis and learning elements are discussed, the control element must be developed. Recall that the control element is a basic FC, and is constructed based on the fundamentals of FCs set forth in Part I of this book.

14.3.1 Control Element

The FC used rules of the form:

IF <***condition***> THEN <***action***>

where the ***condition*** portion of the rules included information concerning the current status of the flow rates of ammonia, formaldehyde, and water (q_A, q_F, and q_W) and the temperature in the tank (T). The ***action*** portion of the rules contain instructions for adjusting q_A and q_W. Notice that q_A and q_F appear in both the condition and the action portions of the rule. This is not a typographical error; the values of these flow rates are the controls and thus appear in the action portion of the rule. However, since there are limitations on the amount these values can change, the current flow rates affect the decision that is ultimately made, and thus appear in the condition portion of the rules.

Five linguistic terms were used to characterize T, and q_A, while four were used for q_F, and three were used for q_w. The action variables, q_A and q_w, were characterized by seven linguistic terms each. As a consequence of the above decelerations, the FC was composed of 300 rules (5 * 5 * 4 * 3 = 300), a result of the inclusion of all possible combinations of the condition variables as described by the chosen linguistic variables. A standard center of area method was used for defuzzification.

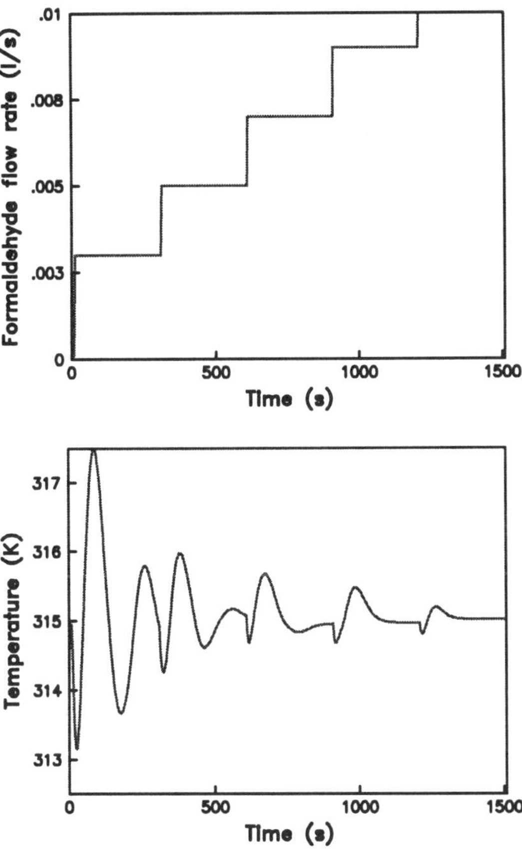

Figure 14.2 - The control element can be used independently of the analysis and learning elements to effectively control the hexamine system when the process dynamics are not changing.

The non-adaptive hexamine FC produced using the above guidelines can be used without adaptive capabilities to effectively control the hexamine system — until the system becomes dynamic, i.e., until the concentrations of the reactants are altered. Figure 14.2 shows the performance of the FC for a situation in which the hexamine system is not experiencing changing system dynamics. Certainly, the FC of the control element performs well when the process dynamics remain constant. However, when the system dynamics are altered (e.g., when the concentrations of the reactants are altered) the performance of the non-adaptive controller is not optimum, because the non-adaptive FC has no mechanism to accommodate the changes in system response caused by the changing concentrations.

14.3.2 Analysis Element

The changing system dynamics in the hexamine system can be accounted for by allowing the FC to alter its conception of the terms used in its rule set, or by changing the rules themselves. This is accomplished by altering the membership functions associated with the action and control variables in response to changes in the hexamine environment. Here, this task is accomplished with analysis and learning elements.

The analysis element is charged with the task of identifying when and to what extent the process dynamics change. As seen in the previous chapter on the pH system, this can be accomplished by employing a GA to tune the parameters of a computational model of the physical system. The parameters are adjusted so that the response of the simulated system matches the response of the actual system. In the current hexamine system, the coding scheme and the fitness function required by the GA to successfully complete the curve-fitting problem are the same as were set forth in the pH system of Chapter 14. The computational model that was presented in the previous section performed acceptably.

14.3.3 Learning Element

The analysis element notifies the learning element of the nature and extent of the changes occurring in the hexamine system. Based on this information, the learning element employs a GA to alter either the membership functions or the rules used in the FC of the control element. To do this successfully, the two issues of coding and fitness function must be addressed.

The parameters that have to be coded in the quest for an efficient set of fuzzy membership functions and a viable rule set are points that define all of the membership functions (the linguistic terms) and the real-valued consequents associated with the rules. The tasks of both tuning the membership functions and selecting the rules have been addressed in earlier chapters of the book. As before, concatenated, mapped, unsigned binary coding was used in this application. In this coding, each portion of the bit strings represented either one of the four points needed to define a trapezoidal fuzzy membership function, or a real-value consequent associated with one of the rules. Since there are 31 trapezoidal membership functions needed to characterize all of the variables considered in the FC of the control element, there are 124 parameters that must be represented to completely define the membership functions. However, due to the symmetry present in the problem, only 62 parameters had to be defined to completely describe a membership function set. Since there were 300 rules in the FC of the control element, there were an additional 300 parameters to represent. The 362 parameters were each allotted four bits, resulting in bit-strings of length 1448.

Now that an appropriate coding has been determined, a second issue must be addressed: how to evaluate the merit of each string (each membership function set). This task of defining a fitness function is always application specific; it always comes down to accurately describing the goal of the controller. In this case, the objective of the controller is to drive the hexamine to the desired temperature setpoint in the shortest time possible and to maintain that desired temperature. Thus, it would appear that the fitness function of the pH system described in Chapter 14 is again applicable. However, we would also like for the hexamine FC to maximize the amount of hexamine being produced without wasting reactants. After all, this is the way the company makes a profit and stays in business. The ability of the adaptive system to achieve these objectives can be represented by a fitness function that, first and foremost, specifies how well the controller reduces the difference between the current temperature and the desired temperature over some finite time period. Additionally, the fitness function must indicate the amount of hexamine produced. Finally, to ensure that a robust FC is obtained, the fitness function should reflect the controller's ability to reach the setpoint from a number of initial condition cases. Mathematically, this fitness function is expressed as:

$$f = \sum_{t=0\text{ sec}}^{t=1510\text{ sec}} [w_1(T - 315.0) + w_2(0.01 - m_{hex})] + w_3 q_F$$

where f is the cost function, m_{hex} is the total moles of hexamine leaving the tank, $w_1 = 3.0$, $w_2 = 1000.0$, and $w_3 = 100.0$ are weighting function values. The weighting function values are selected so that all three products in the objective function are of the same order of magnitude. In this way, the GA places the same emphasis on all three objectives. The objective function was evaluated over two different forcing functions so that a general purpose adaptive controller would be produced.

It is important to note here that the learning element took approximately two hours to locate an effective control strategy to be used by the control element. The lengthy computational time was due to the amount of information that was being processed — recall that the learning element was searching for effective rules and membership functions encapsulated in strings that were 1448 bits long. However, in the hexamine system, the process dynamics generally do not change quickly. In fact, it is not uncommon for the process dynamics to change only on the order of days. Thus, the computational time required by the learning element is acceptable. However, in other systems (for instance the chaotic ball system of Chapter 17), the time necessary to locate effective rules in real time is simply not available, and other alternatives must be considered.

14.4 RESULTS

An adaptive hexamine GA-FC was produced for the hexamine system. Each time a change was detected in the hexamine system by the analysis element, a GA was used by the learning element to locate a new, more efficient rule set and membership functions. The performance of this adaptive hexamine GA-FC is demonstrated for a situation in which the concentration of one of the reactants is changing.

Consider the situation where the concentration of a reactant is changing due to an external agent. The adaptive hexamine GA-FC alters the rules and membership functions it uses to enact its production rules (which do not change) because the process dynamics are altered when the concentration changes. Figure 14.3 compares the performance of the adaptive GA-FC with the non-adaptive FC of the control element that has been disconnected from the analysis and learning elements. According to Figure 14.3(b), both the non-adaptive and the adaptive controllers do a reasonably good job of maintaining the temperature in the reactor. However, the adaptive GA-FC produces more hexamine (the adaptive GA-FL produces 0.81 moles of hexamine per mole of ammonia while the non-adaptive FC produces only 0.65 moles of hexamine per mole of ammonia). The adaptive controller outperforms the non-adaptive FC because the adaptive controller is flexible enough to accommodate the changing process dynamics.

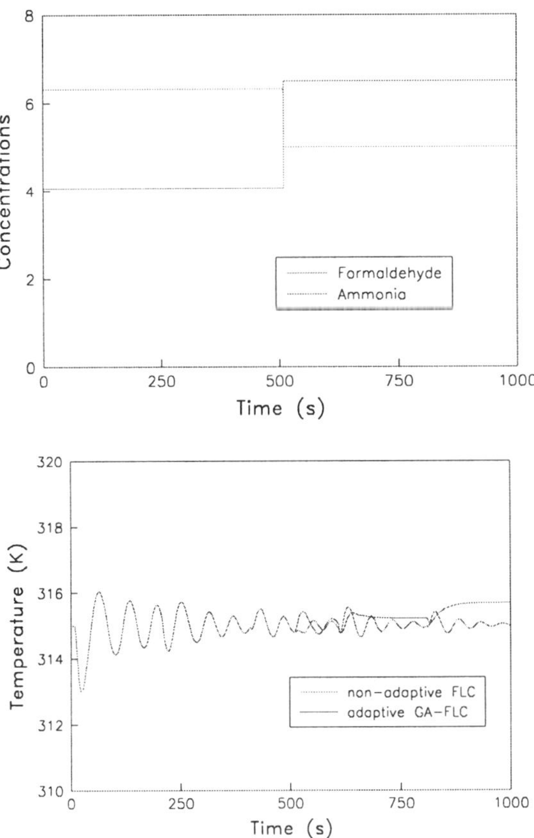

Figure 14.3 - (a) The external agent alters the concentration of a reactant. (b) The adaptive GA-FC was able to drive the system temperature to the setpoint while producing more hexamine than the non-adaptive FC.

The results presented in this section demonstrate much of the power of the adaptive FCs possible through the implementation of the general purpose software architecture defined in Chapter 14. The adaptive controller is able to maintain a high degree of control over the hexamine system despite drastic changes in the system characteristics. Perhaps the most intriguing aspect of this application is the fact that the system parameters that were altered do not appear in the rule set the FC employs. All changes in FC performance are due to alterations in the membership functions. Based on the results presented, adaptive GA-FCs will allow the volatile characteristics of many industrial chemical systems to be controlled via on-line changes to the membership functions.

14.5 REVIEW/PREVIEW

The architecture for achieving adaptive control set forth in Chapter 13 was applied to a different chemical system. It was used to maintain the temperature in a system in which an exothermic chemical reaction was taking place. The adaptive controller is based on three specific elements: (1) a control element, (2) an analysis element, and (3) a learning element.

The control element consists of an FC and is responsible for manipulating the hexamine system. The analysis element recognizes when and to what extent the hexamine system's process dynamics change. The analysis element employs a computer model of the hexamine system and a GA. The learning element uses the results of the analysis element to update the control strategy of the control element. The learning element employs a fuzzy controller and a genetic algorithm.

The results provided in this Chapter suggest that the approach to adaptive control introduced in Chapter 13 is robust. Further, the results presented here demonstrate the capability of the approach to incorporate additional factors into the adaptation of the control system. The multi-purpose objective considered here is similar to those generally considered in industry. Furthermore, few traditional control systems are able to simultaneously consider numerous control objectives.

This chapter and the one immediately before it were used to introduce the idea of adaptive process control. In the next chapter we will present an application of the adaptive control system to a process from the minerals industry. Then, in the remainder of the book we will consider facets of the GA-FC combination that we have not yet addressed in this book.

REFERENCES

Fogler, H. S. *Elements of Chemical Reaction Engineering*. Prentice-Hall, Englewood Cliffs, NJ (1986).

Goldberg, D. E. *Genetic Algorithms in Search, Optimization, and Machine Learning*. Addison-Wesley, Reading, MA (1989).

Holland, J. H. *Adaptation in Natural and Artificial Systems*. The University of Michigan Press, Ann Arbor, MI (1975).

Karr, C. L. *Analysis and Optimization of an Air-Injected Hydrocyclone*, PhD Thesis, The University of Alabama, Tuscaloosa, AL (1989).

Karr, C. L. Genetic algorithms for fuzzy logic controllers. *AI Expert*, **6**, (2), 26-33 (1991).

Karr, C. L. and Gentry, E. J. Real-time pH control using fuzzy logic and genetic algorithms. *Proceedings of the Annual Meeting of the SME* (preprint number 92-49), Phoenix, AZ.

Kelly, E. G., and Spottiswood, D. J. *Introduction to Mineral Processing*. John Wiley & Sons, New York (1982).

Kermode, R. I., and Stevens, W. F. Experimental verification of the mathematical model for a continuous stirred-tank reactor. *The Canadian Journal of Chemical Engineering*, **April**, 68-72 (1965).

Chapter 15

Adaptive Control of Column Flotation

15.1 INTRODUCTION

The focus of this chapter is an ongoing effort to develop and implement an adaptive process control system for a column flotation unit. Flotation is a physical process commonly used to achieve mineral separation, i.e., to separate the valuable products from the waste material in an ore sample. Column flotation units, a new type of flotation technology, are rapidly increasing in popularity because of their efficiency and mechanical simplicity (Finch and Doby, 1990). However, due to the complex process dynamics associated with column flotation, it has been difficult to achieve more than stabilizing control in which the system parameters are driven to some predetermined setpoints.

The adaptive control system described in the previous two chapters is an inviting alternative to the "man-in-the-loop" control currently used to control column flotation. To be consistent with the presentations in Chapters 14 and 15, the current chapter will proceed as follows. The column flotation system is described. Then, the details of applying the individual elements of the adaptive control system to the column flotation unit are presented.

15.2 COLUMN FLOTATION SYSTEM

Valuable mineral commodities are found as ores. These ores vary considerably in composition. Generally, the individual mineral grains of the ore are locked tightly together and therefore the ore must be ground to liberate the desired mineral from the undesired minerals called gangue. The amount of size reduction required to achieve liberation via grinding is variable, thus the resulting sample is variable in both composition and size. The desired mineral contained in the sample must be separated from the gangue using techniques such as flotation. Since the feed sample is so varied, the control problems associated with separation methods such as flotation can be quite difficult, and can benefit from adaptive control systems.

Flotation is an established method for concentrating minerals (Sastry, 1978). Separation occurs when selected particles are rendered hydrophobic (water-repellent) with chemical agents (called collectors) and are attached by chemical and physical forces to air bubbles moving upward through the slurry. Hydrophilic (water-attracted) particles are not attached to the air bubbles and settle out separately. Conventional flotation cells have been a cornerstone of the mineral processing industry for a number of years. In these conventional flotation cells ore particle attachment occurs in a turbulent intensely stirred liquid containing air bubbles and the ground ore. The bubbles with their attached ore minerals rise to the surface of the vessel to form a froth. The froth is removed and the valuable ore mineral reclaimed from the froth. However, column flotation is becoming increasingly popular as an alternative to conventional flotation cells because of its efficiency and mechanical simplicity. Several copper-molybdenum recovery operations in the United States and Canada are currently operating flotation columns while numerous other companies are running pilot-plant tests (Agar, Huls, and Hyma, 1991). The popularity of column flotation separators should continue to grow as a better understanding of the underlying mechanics of the operation is gained.

Figure 15.1 shows a schematic of a column flotation unit. A chemically treated mineral slurry (feed) is fed into the column approximately one-third of the way down from the top of the column. Bubbles are introduced at the bottom of the column. As the feed particles sink and the bubbles rise, collisions occur. As they do in conventional flotation, hydrophobic particle species attach to the bubbles and rise to the top, while hydrophilic species are repelled by the air are not affected and sink to the bottom; thus, again a separation occurs.

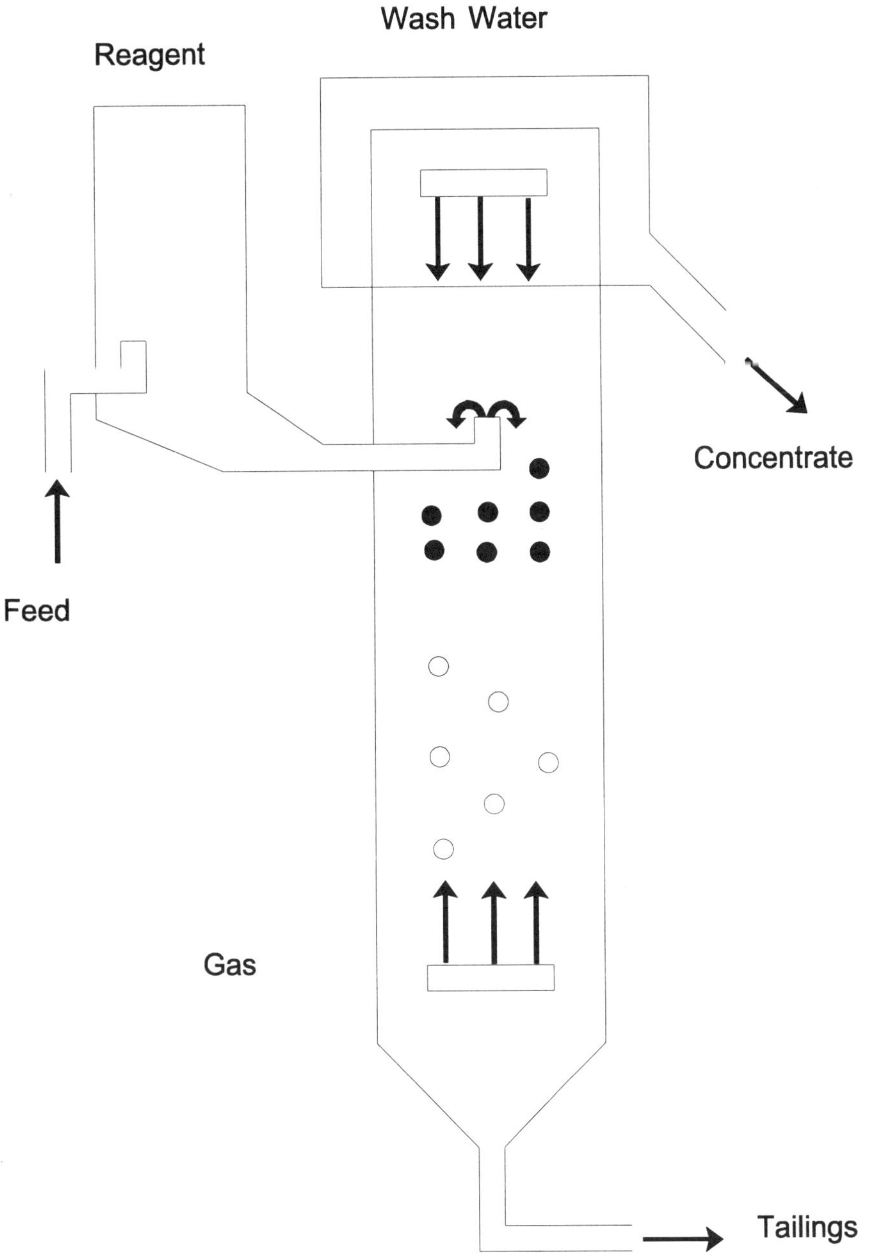

Figure 15.1 - Column flotation units are gaining popularity because of their simple design and their effectiveness in separating a variety of minerals.

The effectiveness of mineral separations is characterized by grade and recovery. Grade is a measure of the quality of the separation; it is a measure of the purity of the product. Recovery is the percentage of mineral fed into the separation process that is recovered in the product. For instance, say we are separating a feed material that contains 100 grams of coal and 50 grams of ash. Upon completion of separation, we have a product that contains 90 grams of coal and 20 grams of ash (the rest of the coal and ash are removed as tailings). The separation recovery is the percentage of coal in the original feed that appears in the product, i.e. 90/100 = 90%. The grade is the relative purity of the product, i.e. 90/110 = 81.8%. When separating mineral samples, both grade and recovery must be considered simultaneously.

Although the mechanics of column flotation are not difficult, and column flotation units offer distinct advantages for improving the metallurgical efficiency of a flotation circuit, there is room for improvement in the control of column flotation. Process variables (also known as condition variables) such as pulp level (the height of the froth in the column), gas hold-up, bias rate (the net flow into the column), bubble size, feed slurry rate, and overflow rate must constantly be measured and monitored if maximum performance is to be achieved. Furthermore, additional condition variables such as grade, recovery, and system capacity must be closely monitored so that controlled or action variables such as flow rates, pH, aeration rate, and chemical agent addition can be adjusted to provide for optimum process efficiency. The objective of the current research effort is to produce a system that is capable of efficiently controlling a column flotation unit in an industrial environment.

As will be pointed out in the next section, the architecture presented in the previous two chapters for achieving adaptive process control requires a computer model of the problem environment. Thus, this approach is actually a form of *model following control* where the computer model is used for system characterization. After the parameters of the problem environment are accurately identified, the strategy used for manipulating the column flotation unit is altered or adapted to provide optimum performance. Without question, a vital component of the process control architecture is a computer model of the column flotation unit. In recent years, numerous researchers have worked to develop an accurate model of column flotation. Some of the more noteworthy research has been performed by Finch and Doby (1990), Herbst and Rajamani (1989), Luttrell, Adel, and Yoon (1987), Sastry and Lofftus (1988), and Ynchausti, Herbst, and Hales (1988). Despite these efforts to develop a computer model of column flotation, there is still no single model that is suitable for the control system that is being proposed because each of the above models has a weakness that could seriously handicap the adaptive controller. Therefore, the current research effort includes the development of a neural network model of a column flotation unit. Previous efforts have demonstrated the ability of neural networks to accurately model conventional flotation units (Karr, Gentry, and Stanley, 1995), and therefore should be effective in developing model following controllers for column flotation units. The process control architecture that is outlined in the next section should provide a platform on which an efficient control system can be built.

15.3 STRUCTURE OF THE ADAPTIVE CONTROLLER

For convenience Figure 14.2 is repeated here as Figure 15.2. This figure shows a schematic of the adaptive process control system architecture. The heart of this control system is the loop consisting of the control element and the problem environment (the column flotation unit). The control element receives information from sensors in the problem environment concerning the status of the ***condition variables***, i.e., grade, recovery, system capacity, etc. It then computes a desirable state for a set of ***action variables***, i.e., wash water rate, gas rate, chemical agent addition, etc. These changes in the manipulated variables force the problem environment toward a predetermined setpoint. This is the basic approach adopted for the design of most closed loop control systems and, in the form described above, includes no mechanism for adaptive control.

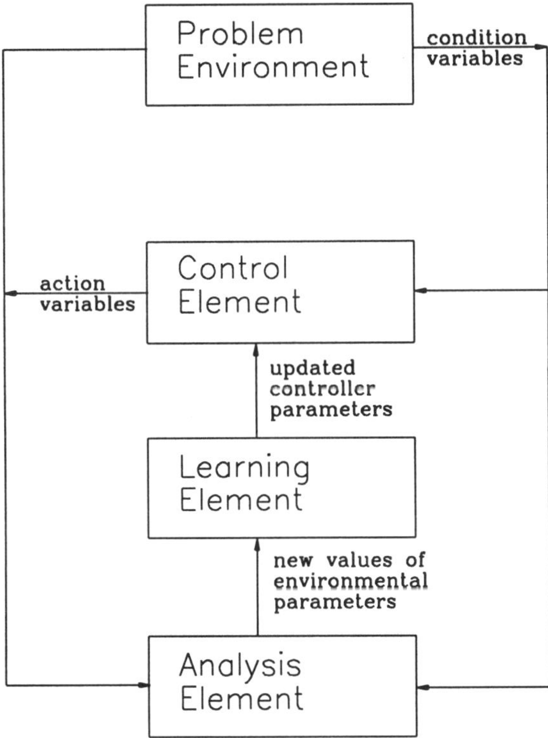

Figure 15.2 - The adaptive control system architecture is robust enough to be applicable to a variety of industrial systems including a column flotation system.

The adaptive column flotation control system presented here is dependent on a neural network model of column flotation. In the remainder of this section, a summary of the three requisite elements of the controller is provided. Special attention is focused on the analysis element since it is dependent on the neural network model. Recall that it is the analysis element that is most dependent on the computer model of the system being controlled.

15.3.1 Control Element

The control element receives feedback from sensors in the column flotation unit, and based on the current state of the condition variables, must prescribe appropriate values of the action variables. The control element in the current system is an FC. A valid (yet simple) rule for an FC used to manipulate a column flotation unit is:

IF {GRADE is **LOW**} THEN {AIR FLOW RATE is **INCREASED**}.

This rule says that if the grade of the product being produced by the column flotation unit is lower than desired, then the air flow rate should be increased. The fuzzy terms (**LOW** and **INCREASED**) are defined by triangular membership functions. Although the way in which a column flotation unit responds to changes in the operating parameters is not yet fully understood, there have been some strides made in this area. In fact, the basic form of a process matrix for the response of column flotation units has been established (Finch and Doby, 1990). Table 15.1 represents a summary of the current level of understanding of the response of column flotation units. Note that the table summarizes the response of the controlled variables caused by an increase in the manipulated variable (I indicates that the controlled variable increases with an increase in the manipulated variable, while D indicates that the controlled variable decreases with an increase in the manipulated variable). This table will serve as the cornerstone of a column flotation FC in that the rules for an FC are written based on the logic included in the process matrix. Unfortunately, there are situations for which the process matrix is not as accurate as one would like. However, the adaptive process control architecture includes a mechanism for dealing with such situations.

Table 15.1 - Process Matrix for Column Flotation

CONDITIONS	ACTIONS			
	Bias Rate	Gas Hold Up	Grade	Recovery
Wash Water Rate	Increase	Increase	Increase	Decrease
Gas Rate	Decrease	Increase	Decrease	Increase
Frother	Decrease	Increase	Decrease	Increase
Feed Rate	Decrease	Increase	Increase	Decrease

15.3.2 Analysis Element

The analysis element recognizes changes in parameters associated with the problem environment not taken into account by the rules used in the control element. In the column flotation unit, the main parameter of concern is the composition of the feed into the column. Changes in the concentration of the feed can dramatically alter the way in which the column flotation unit responds to changes in the manipulated variables; thus, forming a new problem environment requiring an altered control strategy. Recall that the FC used for the control element will not include the composition of the feed in the rules it uses to manipulate the column flotation unit. Therefore, some mechanism for altering the prescribed actions must be included in the control system. But before the control element can be altered, the control system must recognize that the problem environment has changed, and compute the nature and magnitude of the changes.

The analysis element recognizes changes in the system parameters by comparing the response of the physical system to the response of a model of the column flotation unit. A neural network model of column flotation has been developed and tested for this purpose. The neural network has four input nodes, six middle nodes, and two output nodes. It computes the grade and recovery of a column flotation unit based on the following parameters: (1) flow rate of the slurry into the column, (2) percent solids in the feed slurry, (3) amount of collector, and (4) the air flow rate. A schematic of the neural network model is shown in Figure 15.3.

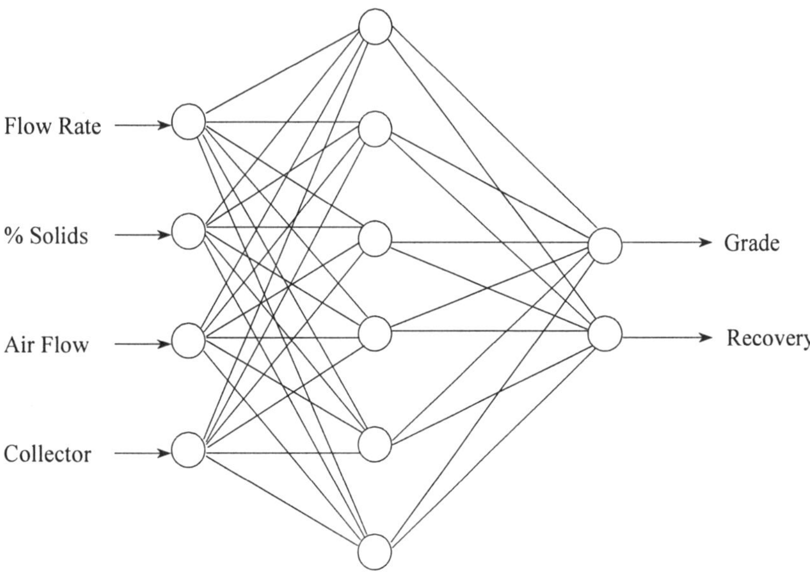

Figure 15.3 - The neural network model of column flotation consists of four input nodes, six middle nodes, and two output nodes. Its accurate calculation of grade and recovery allows the analysis element to effectively determine when and to what extent the column flotation system has changed.

Figure 15.4 shows the effectiveness of the neural network model of column flotation. This figure shows the neural network model's predicted response for a column flotation unit. Notice that the neural network model predicts accurately except in the extreme cases.

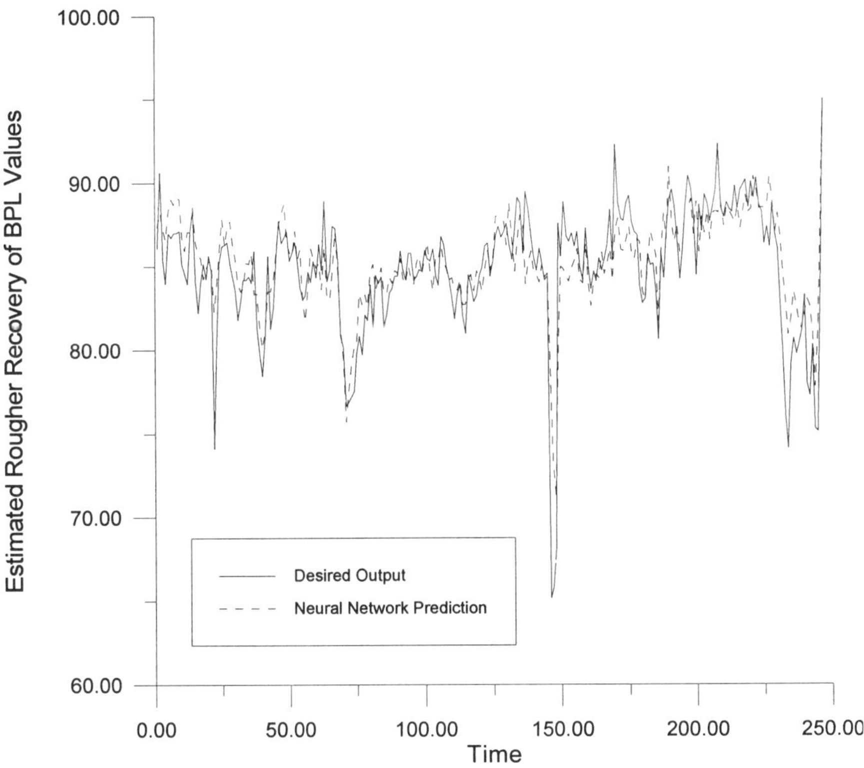

Figure 15.4 - The neural network model of column flotation accurately predicts the response of a real-world system.

As demonstrated in the two preceding chapters, once an effective model of the system has been developed, the analysis element's job of determining when and to what extent the column flotation system has changed becomes a curve-fitting problem. The curve-fitting problem is solved using a genetic algorithm for tuning the model parameters as described in earlier chapters. The coding scheme and fitness function employed in the analysis element for the column flotation controller are the same as outlined previously. When the GA has completed its task of locating the correct values of the parameters in the physical system, it has solved a system characterization problem. At this point, enough information will have been computed concerning the column flotation unit that the control strategy can be updated by the learning element.

15.3.3 Learning Element

The learning element alters the control element in response to changes in the problem environment. It does so by altering the rules and membership functions employed by the FC of the control element. Since the parameters in the problem environment that get changed do not appear in the FC rule set, the only way to account for these conditions (outside of completely revamping the rule base) is to alter the membership functions employed by the FC. A learning element that utilizes a GA to locate efficient rules and membership functions used in an FC that controls a column flotation unit has been developed. The architecture is currently in place, but must be modified in accordance with the control and analysis elements.

15.4 REVIEW/PREVIEW

The application of the adaptive control system architecture introduced in Chapter 14 to a column flotation unit was discussed. The control system has not yet been completed. The strategy employed by the adaptive system uses GAs to fashion three components necessary for a robust, comprehensive adaptive process control system: (1) a control element to manipulate the problem environment, (2) an analysis element to recognize changes in the problem environment, and (3) a learning element to adjust the control strategy.

Chapters 13, 14, and 15 of this book provided an indication of the robust nature of the architecture developed for achieving adaptive process control. Additionally, it is the authors' hope that the reader has generated some original ideas on ways to improve the results presented. In the next three chapters of this book, we intend to address some issues we were forced to face while we were developing working systems. In Chapter 16 we will address the issue of fuzzy controllers as "approximate controllers." We will consider the development of an FC for controlling a chaotic system in which achieving precise results are a must. In Chapter 17 we will consider what happens to FCs as the number of condition variables increases. We will consider an application completed for the U.S. Army in which a GA was used to design a FC for helicopter flight control. In Chapter 18 we will consider an alternative to the expert system approach we have presented thus far in the book. We will consider an adaptive system that has grown out of the GA literature called classifier systems. We will introduce a conventional classifier system, and discuss the extension of such a system by allowing for the fuzzification of its rules.

REFERENCES

Agar, G.E., Huls, B. J., & Hyma, D.B., (Eds.) (1991). *Column '91*. Montreal: The Canadian Institute of Mining and Metallurgy and Petroleum.

Finch, J.A., & Doby, G.S. (1990). *Column flotation*. Toronto: Pergamon Press.

Herbst, J.A., & Rajamani, K. (1989). Models for the dynamic optimization of mineral processing plant performance. In K. Sastry and M. Fuerstenau (Eds.). *Challenges in Mineral Processing*, Denver, CO: Society for Mining, Metallurgy, and Exploration, 709-739.

Karr, C.L. (1991). Genetic algorithms for fuzzy logic controllers. *AI Expert*, **6**, 26-33.

Karr, C.L., Gentry, E.J., & Stanley, D. A. (1995). *An adaptive system for process control* (Report of Investigations number 9563). Washington, DC: U.S. Department of the Interior, Bureau of Mines.

Luttrell, G.H., Adel, G.T., & Yoon, R.H. (1987). *Modeling of column flotation.* Society for Mining, Metallurgy, and Exploration Annual Meeting, Preprint number 87-130.

Sastry, K.V.S. (1978). *Beneficiation of mineral fines, problems and research needs*. Report of a workshop held at Sterling Forest, New York, 78-84.

Sastry, K.V.S., & Lofftus, K. (1988). Mathematical modeling and computer simulation of column flotation. *Column Flotation '88*, Denver, CO: Society for Mining, Metallurgy, and Exploration, 57-63.

Ynchausti, R.A., Herbst, J.A., & Hales, L. B. (1988). Unique problems and opportunities associated with automation of column flotation cells. *Column Flotation '88*, Denver, CO: Society for Mining, Metallurgy, and Exploration, 27-33.

Zadeh, L.A. (1973). Outline of a new approach to the analysis of complex systems and decision processes. *IEEE Transactions on Systems, Man, and Cybernetics*, **SMC-3**, 28-44.

Chapter 16

Control of a Chaotic System

16.1 INTRODUCTION

In our development of adaptive controllers for complex systems we have: (1) introduced the basic concepts of fuzzy systems, (2) discussed the adaptive capabilities of genetic algorithms, (3) provided techniques for modeling complex systems, and (4) tied these ideas together to develop adaptive, model-following controllers. There are issues and questions that remain about adaptive fuzzy systems. First, are fuzzy control systems that use approximate reasoning capable of providing more than approximate results? Second, since the systems considered in this book until now have had a limited number of condition variables, are fuzzy controllers capable of controlling systems in which there are large numbers of conditions variables? Third, are there other ways to design adaptive controllers?

The current chapter considers the first issue above, accuracy. The next two chapters address the concerns associated with the second and third issues. Chapter 17 presents the design of a helicopter flight controller. This effort demonstrates techniques for effectively managing large numbers of condition variables. Chapter 18 presents a fuzzy classifier system that provides an alternative approach to developing adaptive systems.

In the current chapter we investigate the possibility that fuzzy systems provide only approximate solutions. The vehicle for our effort is the design of a genetic algorithm-tuned fuzzy controller for a chaotic system. Chaotic systems are highly sensitive to control actions taken on them, and require extremely accurate controllers. The controller developed in this chapter is capable of controlling a chaotic system that is difficult to regulate, yet the system is simple enough for the reader to readily grasp the underlying mechanics of the system.

16.2 CHAOTIC SYSTEMS

Researchers have been aware of chaotic behavior in engineering systems for a number of years (Poincaré, 1921; Van der Pol and Van der Mark, 1927) —

generally they simply termed such systems unstable and avoided operating in the chaotic regions. However, in recent years new phenomena that reveal an underlying order in chaotic systems have been discovered in nonlinear dynamics. Exploiting this underlying order has permitted prediction and control of chaotic systems.

Chaos, unfortunately, is not easily defined in either nonscientific or scientific terms. In nonscientific terms chaos is generally associated with a complex system that is without apparent pattern. In scientific terms these systems are no easier to define, but are generally characterized by three tangible qualities: (1) they act irregularly, i.e., they are not repetitive, (2) they are extremely sensitive to initial conditions, and (3) their behavior is described by mathematical equations that are often quite simple. Chaotic systems have been identified in a number of scientific fields including electrical engineering (Moon, 1987), hydrodynamics (Berge, Pomeau, and Vidal, 1984), chemistry (Borman, 1991), and mechanical engineering (Tufillaro and Albano, 1986).

While at the U.S. Bureau of Mines, we developed an interest in chaos because many mining and metallurgical systems undergo dramatic changes with small perturbations in boundary conditions. Examples of this chaotic phenomena have been observed in electric arc furnaces (Ochs and Hartman, 1988), the fracture of materials (Brady, 1978), and instabilities of mine structures (Coughlin and Kranz, 1991). Uncontrolled complex systems can prove to be expensive or dangerous. Thus, the ability to successfully predict and control chaotic systems can represent tremendous economic savings and substantial improvements in safety for the minerals industry, and for numerous other industries.

Researchers have recently begun to investigate techniques for efficiently controlling chaotic systems. One of the first and most important efforts in this area was made by Ott and his co-workers (Ott, Grebogi, and Yorke, 1990) who utilized a technique called delay coordinate embedding. This technique is highly mathematical and requires an in-depth understanding of the physical system's response to external stimulus. Shortly thereafter, Romeiras and his co-workers (Romeiras, Ott, Grebogi, and Dayawansa, 1991) developed a method of controlling chaos by applying a small control that proved to be effective on a small scale.

Control of the chaotic system addressed herein, like virtually any control problem with two input variables, can be viewed as a surface-fitting problem; an appropriate control action is simply a function of the state of the two variables in the problem environment. Both fuzzy logic (Cao, Kandel, and Li, 1991) and GAs (Karr, Stanley, and Scheiner, 1991) have been used to solve curve-fitting problems. Thus, a precedent exists for utilizing a GA-FC for the task of defining a control surface using data from a problem environment. Control surfaces for chaotic systems are generally very difficult to obtain, and once obtained, they are quite complex.

In this chapter, a particular chaotic system is presented for which the control surface has been obtained. Specifically, a ball bouncing on an oscillating table is considered. The frequency of oscillation of the table is adjusted so that the ball bounces to a constant height. The control variable, the desired frequency of oscillation, ω, is a function of two state variables: (1) the velocity of the ball when it leaves the table, v, and (2) the phase angle of the table at impact, θ.

The control surface of the chaotic bouncing ball system is very complex. However, despite its complexity, this surface can be defined accurately with fuzzy production rules. A GA is used to specify the fuzzy membership functions because the FC must be extremely accurate to suitably control the chaotic system. Unfortunately, just using GAs to define the membership functions does not guarantee the necessary precision. The FC must utilize a large number of membership function forms, and it must also be provided with great flexibility in defining the consequent for the fuzzy rules. Triangular, trapezoidal, sinusoidal, and exponential membership functions are necessary. The flexibility in defining the membership functions for the consequent of the rules is obtained by employing a *real-valued consequent.*

Details are provided in this chapter for an FC that is then used to control a computer simulation of the chaotic bouncing ball system. The results demonstrate the effectiveness of using an FC to control a chaotic system. These results are important for two reasons. First, controlling chaotic systems is very difficult, and second, fuzzy logic provides a viable alternative to some of the highly mathematical techniques that have been reported. The successful use of fuzzy logic to control a chaotic system demonstrates that inexact reasoning can be used to obtain very precise control.

16.3 THE CHAOTIC SYSTEM

The bouncing ball system was selected simply to demonstrate the effectiveness of using a particular control strategy to manipulate chaotic systems. It was chosen from the many possible chaotic systems because it is easy for most people to gain a "feel" for the response of this system, and because it is relatively simple.

A ball is dropped from an arbitrary height, and bounces on a table that moves sinusoidally up and down as shown in Figure 16.1. While the gravitational force of the system and the coefficient of restitution of the ball are constant, the height to which the ball bounces is determined strictly by the velocity of the ball relative to the table at impact, and by the phase angle of the table when an impact occurs. If the ball impacts the table when the table is moving upward, the relative velocity is large and the ball bounces higher than it would on a stationary table. If the ball impacts the table when the table is moving downward, the relative velocity is small and the ball bounces lower than it would on a stationary table. The net effect is that the height to which the ball bounces is deterministic, yet it has no apparent long-term pattern. Furthermore, the time-history of the ball (its position at a particular time) is extremely sensitive to the initial height from which it is dropped — the system is chaotic.

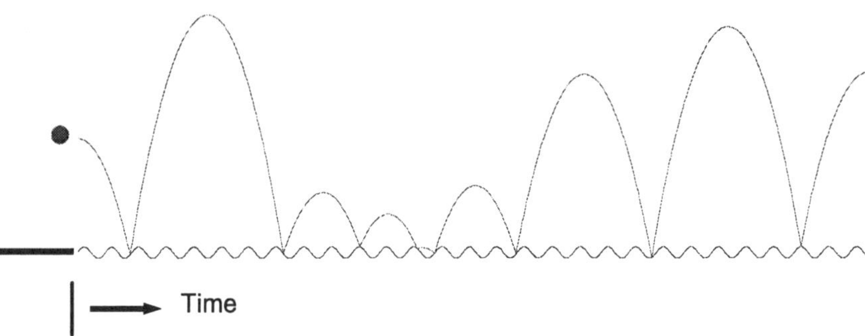

Figure 16.1 - The chaotic bouncing ball system possesses the complex attributes characteristic of chaotic systems yet is simple enough for most people to get a feel for it.

Poincaré maps are used to determine whether a system is chaotic. A Poincaré map for the bouncing ball system appears in Figure 16.2. In this plot, the polar coordinate r, the distance from the origin, represents the velocity of the ball at impact. The angular coordinate θ represents the phase angle of the table when an impact occurs. Thus, a point on the Poincaré map completely describes the state of the bouncing ball system at the time of each collision. The time of the next impact, and the state of the bouncing ball at the time of the next impact can be computed from this information. Since the Poincaré map does not consist of either a finite set of points forming a pattern (indicating the system is in some way repetitive) or a closed orbit (indicating the system is *almost periodic* or *quasiperiodic*), the motion is chaotic. The ball bounces to a seemingly random height each time it contacts the table. However, this does not have to be the case.

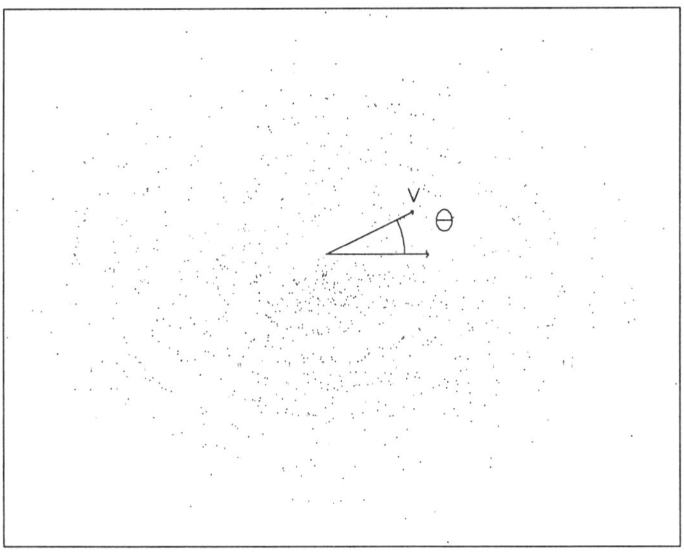

Figure 16.2 - A Poincaré map can be used to determine whether a system is chaotic. In this plot, the polar coordinate **r**, the distance from the origin, represents the velocity of the ball at impact. The angular coordinate θ represents the phase angle of the table when an impact occurs.

The frequency at which the table oscillates is adjustable, and changing this frequency dramatically affects the height to which the ball bounces. In fact, the frequency of oscillation can be altered to force the ball to bounce to a constant height, i.e., to force the ball to always strike the table with the same relative velocity at the same phase angle. Thus, this becomes the objective of a control problem: to adjust the table's frequency of oscillation in such a way as to cause the ball to rise to the same maximum height after each bounce. Moreover, a Poincaré map provides a convenient mechanism for evaluating the effectiveness of a controller. An effective control system should produce a Poincaré map that consists of a single point, because the response of the system will be wholly repetitive.

The bouncing ball system presents an especially appealing problem environment because a control surface can be described analytically using the equations of motion of the system, and thus there is an exact solution to which the control surface developed by the GA-FC can be compared. Second, the details of the GA-FC are described. This description includes the necessary information for utilizing a GA to tune a FC. Third, results are presented that portray the effectiveness of using an FC to manipulate the chaotic bouncing ball system. These results are important because they demonstrate the effectiveness of using fuzzy logic to obtain a precise solution.

16.4 SURFACE-FITTING AND THE ANALYTICAL SOLUTION

As stated above, the problem of controlling the bouncing ball system is a surface-fitting problem: for any possible value of the ball's velocity at impact, **v**, and any possible value of the phase angle of the table, $0 \leq \theta \leq 2\pi$, the frequency of oscillation of the table, ω, can be adjusted to ensure that the ball bounces to the desired height, $y_{desired}$. The equations of motion can be used to compute the control surface. Although the equations of motion for the system are rather simple, obtaining the solution to these equations is tedious. The solution is the following equation, which includes both $y_{desired}$ and the value of ω necessary to ensure the ball bounces to a maximum height of $y_{desired}$:

$$\sqrt{2g[Asin(\omega t) - y_{desired}]} = ev_0 + egt_c + (1 - e)A\omega\cos(\theta) \qquad (16.1)$$

where g is the acceleration due to gravity, A is a constant that indicates the magnitude of the oscillation, e is the coefficient of restitution, v_0 is the initial velocity of the ball, and t_c is the time at which the next collision occurs where $\theta = \omega t_c$.

Equation (16.1) gives the appropriate value of ω for any given values of v_o, θ, e, A, and $y_{desired}$. In this study, the following values were used: e=0.9, A=1.0, and $y_{desired}$=20.0. With these values of the parameters, an appropriate control space consisted of $10.0 \leq v_{ball} \leq 40.0$ and $0 \leq \theta \leq 2\pi$. For values of v_{ball} and θ in this range, equation (5) must be solved numerically. The process of solving this equation, although not difficult using a traditional root-finding method, is quite tedious. There are multiple roots to the equation, and great care must be taken to ensure that the correct root is found and used.

Equation (16.1) was evaluated at numerous points using a combination of a Newton method and a bisection method. This approach allowed for the location and selection of the appropriate roots. The results yield the analytical control surface shown in Figure 16.3. Note the complex nature of this surface. This complexity must be fully described with fuzzy logic, because the bouncing ball system is very sensitive to alterations in ω.

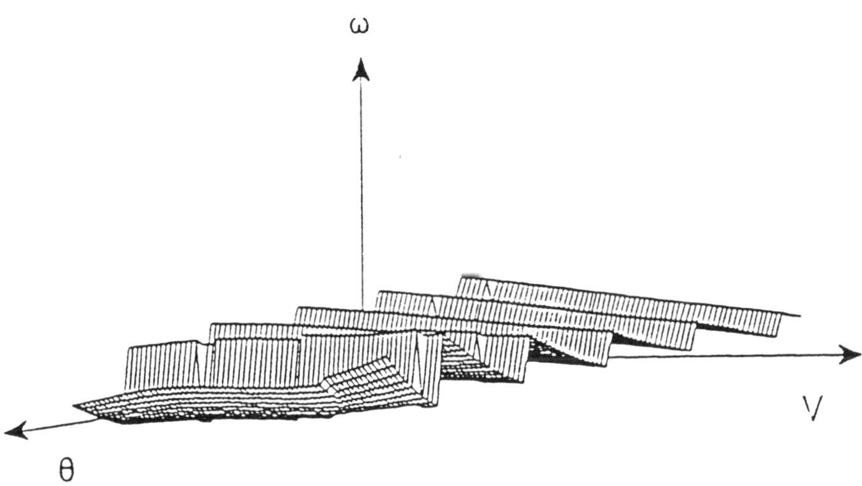

Figure 16.3 - The "analytical control surface" shown above depicts the complex control surface for this chaotic system. Further, since this is a chaotic system, the controller being developed must fit this surface precisely.

16.5 GA-DESIGNED FUZZY LOGIC CONTROLLER

The procedure for developing a GA-FC for the bouncing ball system is the same as presented in the early chapters of this book. Thus, only the details of the application are presented here. As presented here, the chaotic system does not require an adaptive controller because there are no changing system parameters that are not condition variables. An adaptive controller would be required if a parameter such as the coefficient of restitution were changing. However, for our current GA-based approach to work the changes in e would have to occur infrequently relative to the time constant of the system so that the GA would have time to discover new control strategies.

The fuzzy controller takes two input variables, v and θ, and one output variable, ω. Because of the sensitive nature of the chaotic system, numerous linguistic terms are required to adequately represent each of the pertinent controlled variables, v and θ. Twenty-nine linguistic terms were used to represent both v and θ. The manipulated variable, ω, was represented using triangles only, and an infinite number of linguistic terms were used since the real-valued consequent approach was employed. As normal, these linguistic terms are defined by membership functions. The analytical control surface shown in Figure 16.3 depicts the intricate detail of the terrain that must be represented with a fuzzy rule base. Notice that the contours of the control surface seem to be linear in some places, sinusoidal in other places, and exponential in still other places. Those sections of the surface that appear to be linear are accurately represented by triangular membership functions (trapezoidal functions could have also been used). However, to accurately represent the other regions, more complex membership function forms were needed. Intuitively, one might try to represent the sinusoidal regions with sinusoidal membership functions, and the exponential regions with exponential membership functions. Both sinusoidal and exponential membership functions can be represented with appropriate general equations. The sinusoidal membership functions are described with an equation of the following form:

$$f(x) = \frac{1}{2}[\sin(\frac{\pi x}{w} + \frac{\pi}{2} - \frac{c\pi}{w}) + 1.0] \tag{16.2}$$

where c is the center (location of maximum height) and w is the width. The exponential membership functions are described with an equation of the following form:

$$f(x) = 1.0 + \frac{(e^{k*abs(x-c)} - 1.0)}{(1.0 - e^{kw})} \tag{16.3}$$

where c is the center (location of maximum height), w is the width, and k is a constant defining the curvature of the exponential function. This happens to be a case where intuition is well served. It turns out that the above three membership function forms (triangular, sinusoidal, and exponential) can represent the control surface. These membership function forms are shown in Figure 16.4.

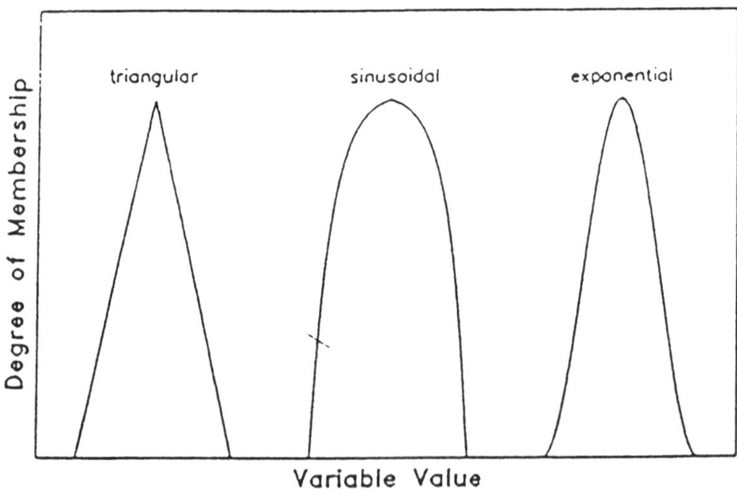

Figure 16.4 - The above membership function forms were necessary and adequate to represent the complex control surface of the chaotic system.

A GA can be used to design both the rules and membership functions for the chaotic FC. Although the problem is larger (29 membership functions for each of the two condition variables) than those in previous chapters, the approach to using the GA is the same. A portion of a bit string is used to represent each of the membership functions while other pieces of the bit string are used to represent rules. Because of the vast size of the search space (there are 29*29=841 rules), the problem was tackled in pieces. The search problem was solved piecewise; only small portions of the search space were solved at a time. In this way, the GA was presented with a manageable problem.

16.6 RESULTS

A GA-FC was developed that can efficiently manipulate the bouncing ball system. This controller alters the frequency of oscillation of the table, based solely on the velocity with which the ball strikes and on the phase angle of the table at the time of impact to ensure the ball bounces to a constant height. As pointed out earlier, the effectiveness of such a controller can be judged from a Poincaré map. Such a Poincaré map for the bouncing ball system wherein a trained GA-FC has been put into operation was created. However, this map included only five points in the space. Thus, the figure was not very clear. There are so few points because the trained controller is put into operation after four bounces. The resulting figure is far different from Figure 16.2, which shows a Poincaré map for the bouncing ball system with no control. The GA-FC has instilled order in an otherwise chaotic system.

For many, a Poincaré map is a natural vehicle for judging the performance of a controller in the realm of chaos. For others, however, it is difficult to use this type of map. To assist interpretation, Figure 16.5 shows the maximum height to which the ball bounces as a function of time. Recall that the objective of the controller was to adjust the frequency of oscillation of the table in such a manner as to force the ball to always bounce to a constant height. Figure 16.5 leaves no doubt as to the effectiveness of the GA-FC. Figure 16.5 shows the same simulation as does the Poincaré map depicted described above. In both, the controller is put into operation after four bounces.

For the simulation depicted in Figure 16.5, the controller is able to adjust the oscillatory frequency of the table so as to immediately cause the ball to bounce to the target height. However, this is certainly not always the case. A limit placed on the range of acceptable values for ω that reflects the limit on the hardware that could be used if the system were to be built. This limitation restricts the controller by restricting the amount of energy that can be imparted to or dissipated from the bouncing ball. As an example, if the ball is bouncing well above the target height, the maximum energy dissipation will occur when the ball strikes the table when the table has its maximum downward velocity. Since the value of ω is restricted, so too is the energy that can be dissipated. Likewise, if the ball is bouncing to a height well below the target height, the maximum energy that can be imparted to the ball will occur when the ball strikes the table when the table has its maximum upward velocity. Since the value of ω is restricted, so also is the energy that can be imparted to the ball.

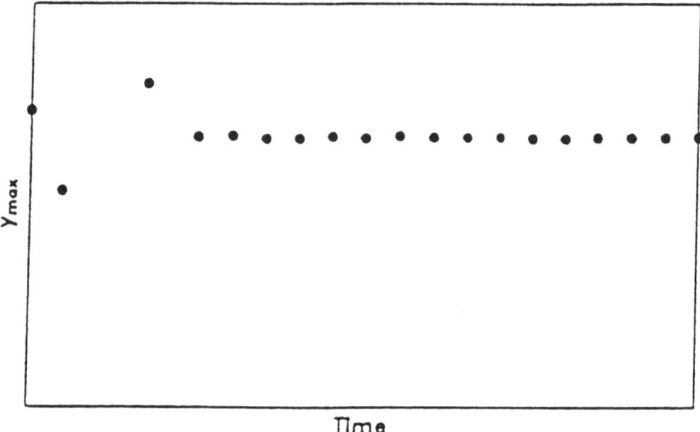

Figure 16.5 - The above plot of maximum ball height as a function of time indicates that the GA-FC effectively manipulates the chaotic system.

The GA-FC produces a control surface that can be compared with the analytical control surface shown in Figure 16.3. The surface produced by the GA-FC is shown in Figure 16.6. Notice that the GA-FC control surface is characterized by all of the complexities contained in the analytical control surface. However, simply comparing these two surfaces in such an inexact fashion is by no means adequate. Therefore, Figure 16.7 shows an error surface, which is simply the absolute value of the difference between the control actions prescribed by the analytical controller and by the GA-FC. Note that there is excellent agreement between the two controllers for most of the space. However, there are a few places in which the difference between the two controllers is substantial. These regions are characterized by discontinuities in the control surface. The GA-FC is sometimes unable to represent these discontinuities adequately because of the fact that fuzzy logic requires a discretization of the search space. Certainly this discretization is more flexible than the discretization possible with conventional set theory. At this point, the authors yield to the classical trade-off made in virtually all computational work; that of accuracy versus computational time. The magnitude of the error surface could well be reduced with greater discretization (the definition of more fuzzy sets), but this adds to the complexity of the controller and increases the time the controller needs to prescribe a control action.

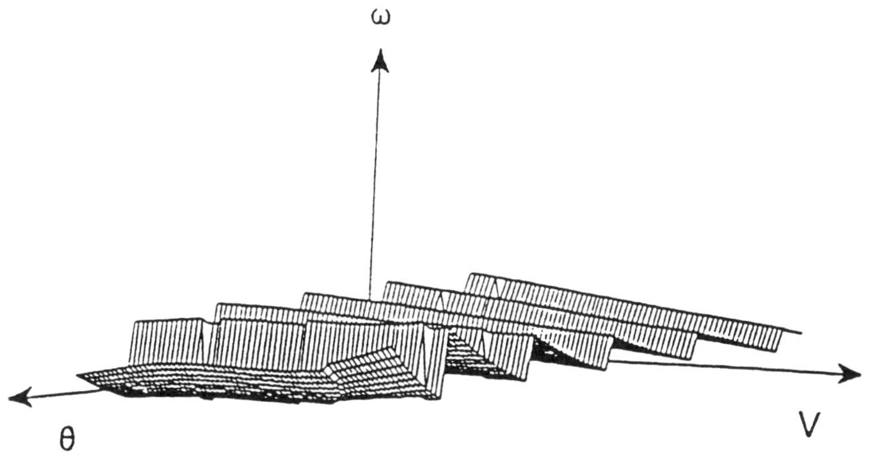

Figure 16.6 - The control surface produced by the GA-FC is remarkably similar to the analytical control surface.

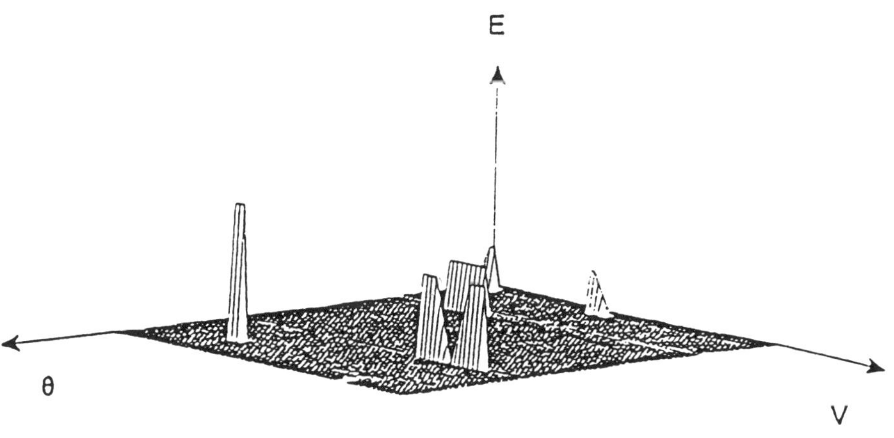

Figure 16.7 - The above error surface indicates the effectiveness of the GA-FC for controlling the chaotic system.

For most people, the information depicted by Figures 16.5 and 16.7 should be adequate to demonstrate the effectiveness of the GA-FC. However, there are some who prefer statistical analysis over graphical comparisons. For these people, the following analysis is included. For the purposes of this analysis, error, *E*, is defined as:

$$E = \frac{\omega_{analytical} - \omega_{GA\text{-}FLC}}{\omega_{analytical}} * 100\% \tag{16.4}$$

where $\omega_{analytical}$ is the oscillatory frequency of the table prescribed by the analytical controller and $\omega_{GA\text{-}FC}$ is the oscillatory frequency of the table prescribed by the GA-FC. One hundred thousand points were sampled to compute the average error, E_{avg}=0.4673%, and the standard deviation, σ=3.141. Moreover, 98.54% of the points sampled had an average error of less than 1.0%. The GA-FC produces an accurate depiction of the analytical control surface.

The above discussion demonstrates the accuracy of the GA-FC for the bouncing ball system. As with most chaotic systems, a controller for the bouncing ball system must be more than just accurate; it must also be fast. It is quite important to note that the analytical controller described earlier in this chapter takes on the order of one second to compute its control action, whereas the GA-FC takes on the order of milliseconds to compute its control action. This requirement for fast and accurate controllers becomes even more important in some other chaotic systems. An example of a chaotic system in which speed is at a premium appears in the steel industry in the form of an electric arc furnace (Ochs and Hartman, 1988).

16.7 REVIEW/PREVIEW

In this chapter we addressed the concern that fuzzy controllers are capable of providing only approximate solutions. A chaotic system requiring precise control actions has been effectively manipulated using a genetic algorithm-designed fuzzy controller. Although the particular application presented in this chapter required a large number of linguistic descriptive terms (29 for each of the condition variables), it is clear that fuzzy systems can provide precise control.

In the next chapter we turn our attention to another issue that should be considered when employing a fuzzy controller. The large number of linguistic terms used in the current chapter to describe the condition variables presented only a small problem. However, a bigger problem occurs when a large number of condition variables must be considered. In such a case, the size of the rule set can rapidly become unmanageable. Such a case occurred for the authors when we were asked by the U.S. Army to investigate helicopter flight-control problems. In the next chapter we will consider ways of exploiting domain knowledge to ease the development of fuzzy controllers; we will discuss the development of a flight controller for a UH-1 helicopter.

REFERENCES

Berge, P., Pomeau, Y., & Vidal, C. (1984). *Order within chaos*. New York, NY: John Wiley & Sons.

Borman, S. (1991). Researchers find order, beauty in chaotic chemical systems. *Chemical and Engineering News*, January, 18-29.

Brady, B. T. (1978). Prediction of failures in mines — An overview. *Bureau of Mines Report of Investigations* (number 8285).

Cao, Z., Kandel, A., & Li, L. (1991). A new model of fuzzy reasoning. In *Fuzzy Sets and Systems*, 311-325.

Coughlin, J., & Kranz, R. (1991). New approaches to studying rock burst associated seismicity in mines. *Proceedings of the 32nd U.S. Rock Mechanics Symposium*, 105-109.

Karr, C. L., Stanley, D. A., & Scheiner, B. J. (1991). A genetic algorithm applied to least squares curve fitting. *Bureau of Mines Report of Investigations* (number 9339).

Moon, F. C. (1987). *Chaotic vibrations*. New York, NY: John Wiley & Sons.

Ochs, T. L., & Hartman, A. D.(1988). Improved arc stability in electric arc furnace steel making. In The U.S. Bureau of Mines' (Eds): *New Steel-Making Technology from the Bureau of Mines*, Bureau of Mines Information Circular number 9195, 2-11.

Ott, E., Grebogi, C., & Yorke, J. A. (1990). Controlling chaos. *Physical Review Letters*, **64** (11), 1196-1199.

Poincaré, H. (1921). *The foundation of science: Science and method*. English Translation, New York, NY: The Science Press.

Romeiras, F. J., Ott, E. Grebogi, C., & Dayawansa, W. P. (1991). Controlling chaotic dynamical systems. *Proceedings of the American Control Conference*, 121-130.

Tufillaro, N. B., & Albano, A. M. (1986). Chaotic dynamics of a bouncing ball. *American Journal of Physics*, **54** (10), 939-944.

Van der Pol, B., & Van der Mark, J. (1927). Frequency demultiplication. *Nature*, **120**, 363-364.

Chapter 17

Helicopter Flight Control

17.1 INTRODUCTION

In this chapter we develop a genetic algorithm-tuned, fuzzy controller for a system that has a large number of condition variables. The message of the chapter is that domain knowledge must be exploited when it is available. But, the leap from saying this to actually doing it is often immense. Thus, this chapter provides the reader with methods for limiting the size of a fuzzy controller in what is the most complex control problem addressed in the book – helicopter flight control.

There is tremendous interest in developing automatic control systems for helicopters. The justification for this endeavor ranges from improvements in safety, commercialization, and flexibility. For instance, there are numerous flight missions that place the pilot in danger. Such missions occur both in the military and civilian sectors (fire fighters for instance). In these situations, it is desirable to replace the human pilot with an automatic control system. Additionally, there are some missions that simply preclude the use of human pilots such as when weapons systems are tested by firing at the flight vehicle; obviously, a situation in which un-manned flight is mandatory.

Unfortunately, developing an automatic pilot for a helicopter is a difficult problem for numerous reasons. Helicopter flight control is characterized by inherent instability, a large number of degrees of freedom, a high degree of coupling between the state variables, and a vast number of state variables to be considered. Despite the difficulty in developing an automatic control system, there are presently algorithms in use for attaining automatic helicopter flight control. In fact, many of these methods are quite effective. The major problem with existing methods is that the algorithms are highly aircraft specific; the algorithms have to be completely re-designed for each type of helicopter on which they are to be used. A robust flight control algorithm, one that is both effective and that can be easily adapted for implementation on a variety of aircraft, would save the helicopter industry, weapons manufacturers, and the military large sums of money. In short, such a control system would represent a major step forward in helicopter flight control.

The genetic algorithm-designed adaptive fuzzy control systems of this book are robust and flexible. Thus, fuzzy controllers are potential solutions to the problem of developing a helicopter control system that is easily adapted to various aircraft. And, since efficient helicopter flight control is obviously achieved by human experts, the information needed to construct an effective fuzzy control system can potentially be extracted via interviews with human pilots. Thus, to pursue this avenue, the information and knowledge used by human pilots must be translated into a form that is appropriate for a fuzzy control system. The resulting fuzzy controller could eventually replace human pilots in dangerous missions, and at the same time be flexible enough to be applicable to a variety of aircraft.

One of the key difficulties associated with developing a control system for a helicopter is the complexity of the aircraft dynamics: the state variables are so highly coupled that it is exceedingly difficult to de-couple the system adequately to implement conventional control strategies. Fortunately, fuzzy controllers do not depend on having a de-coupled system. With a fuzzy controller, the tasks necessary to fly the aircraft can be subdivided. Here, a novel approach is adopted in which individual fuzzy controllers (termed "fuzzy blocks" in this chapter) are written to achieve four goals associated with flying the aircraft: (1) the longitudinal velocity goal, (2) the vertical velocity goal, (3) the lateral velocity goal, and (4) the heading goal. This process of dividing the necessary tasks, which produces a unique fuzzy controller architecture, is similar to the way a human pilot flies a helicopter. Perhaps more importantly, the process of dividing the goal tasks reduces the size of the rule set needed in the control system to a point where it is manageable. This is an extremely important point since the size of the rule set grows multiplicatively as the number of "condition" variables is increased. However, this subdivision of tasks does not reduce the size of the rule set to a point that the rules are immediately apparent. Thus, a genetic algorithm is needed for determining appropriate rules.

This chapter describes a successful effort to develop a fuzzy controller for a UH-1H helicopter. The sections that follow provide details of the fuzzy controller architecture and describe the approach used to determine an effective rule set using a genetic algorithm. Results are provided demonstrating the performance of the fuzzy controller on a computer simulation of a UH-1H helicopter, and on a test flight of an actual aircraft. Although the control system has not yet been adapted for other aircraft, the architecture that has been developed appears to be flexible enough to be applicable to a wide variety of helicopters.

17.2 THE PROBLEM

An automated aircraft controller is useful in hazardous flight situations in which it is desirable to manipulate the aircraft without having a pilot present. Since automatic control systems for fixed-wing aircraft are quite prominent, they seem to be a viable solution. Occasionally, however, the aircraft must be maneuvered in tight areas or maintained at a specific position for extended periods of time. In such instances, an unmanned helicopter is the best solution (Sugeno, Griffin, and Bastian, 1993).

Control of a helicopter is much more difficult to achieve than control of a fixed-wing aircraft. This is true mainly because when a pilot releases the controls of a typical fixed-wing aircraft, the aircraft eventually reaches some steady state flight condition – the aircraft is generally stable. The same is not true for a helicopter. Without constant corrective control inputs, a helicopter tends to diverge from steady-state conditions – the aircraft is generally unstable. In addition, the flight dynamics of a helicopter are highly coupled and vary from aircraft to aircraft as well as from one flight regime to another. The coupling is due, in part, to the large gyroscopic moment created by the main rotor. Any forward pitch motion, i.e. "nose-down", in a helicopter results in a corresponding roll to the right. Control-induced aerodynamic effects introduce additional coupling. These coupled effects make helicopters one of the most difficult aircraft to control.

A typical helicopter configuration is shown in the schematic of Figure 17.1. Here, notice that the body of the helicopter is suspended from a main rotor that provides lift. A tail rotor is employed to counteract the torque produced by the reaction of the air against the main rotor. The pilot uses three primary controls to fly the helicopter: (1) the *collective* is used to adjust the pitch of the main rotor blades in order to increase or decrease lift, (2) the *pedals* are used to adjust the pitch of the tail rotor blades, thus producing more or less torque on the aircraft body as needed, and (3) the *cyclic*, for all practical purposes, determines the orientation of the main rotor with respect to the aircraft body. These three main control mechanisms are shown in Figure 17.2.

The main rotor can be thought of as a disk that produces lift. When the cyclic is pushed forward, the disk tilts forward with respect to the aircraft body causing a forward acceleration. When the cyclic is pushed to the right, the disk tilts to the right causing an acceleration to the right. Unfortunately, each control action excites motion on multiple axes. In other words, the controls are highly coupled and, as a result, flying a helicopter is an extremely difficult task.

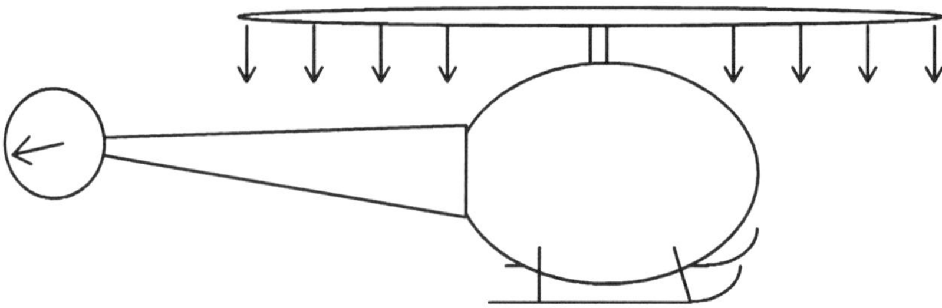

Figure 17.1 - The helicopter produces forces on the surrounding air. These forces are channeled to accomplish flight maneuvers.

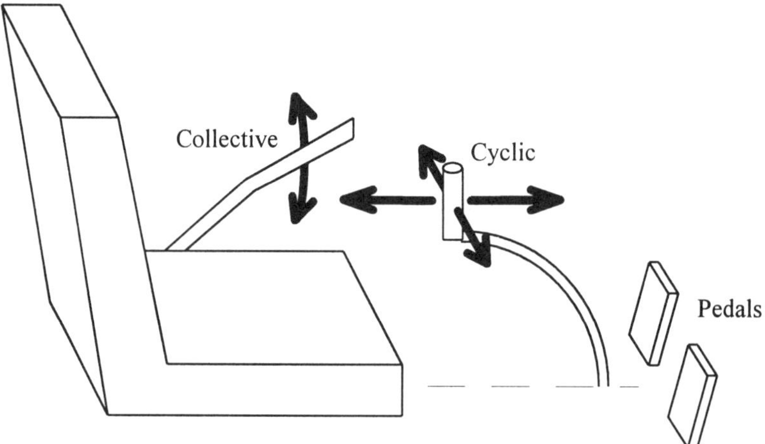

Figure 17.2 - The primary controls in the cockpit of a helicopter include: (1) the *collective* that is used to adjust the pitch of the main rotor blades to increase or decrease lift, (2) the *pedals* that are used to adjust the pitch of the tail rotor blades, thus producing more or less torque on the aircraft body as needed, and (3) the *cyclic*, which for all practical purposes, determines the orientation of the main rotor with respect to the aircraft body.

As a specific example of the coupling effects present in a helicopter, consider a maneuver in which the pilot wants to descend from a hover to some predetermined altitude. To decrease the lift from the main rotor, and thus begin the descent, the collective is pulled downward. This causes the pitch on the main rotor blades to decrease, producing less force on the air. As a result, the counter torque provided by the tail rotor is no longer sufficient to maintain a constant heading. Therefore, the pedals must be adjusted to change the pitch on the tail rotor blades, thereby providing the correct counter torque for the new collective position. Unfortunately, the torque provided by the tail rotor is not generally limited to the vertical axis of the helicopter. In many instances, the tail rotor affects torque about the longitudinal axis resulting in roll. The roll causes a pitch due to gyroscopic precession from the main rotor. Each control produces multiple effects in the motion of the aircraft. As a result, all control actions must be meticulously coordinated in order to perform even the simplest maneuver.

17.3 THE HELICOPTER

Th research effort described in this chapter focuses on the design of a control system for a UH-1H helicopter, sometimes called a "Huey." Initial efforts are focused on testing a fuzzy control system on a simulation of the helicopter. However, the control system is eventually flight tested on an actual aircraft. The control system software was designed using a simulator developed at NASA Ames Research Center. The simulator is a nonlinear, total force and moment model of a single main rotor helicopter. The model has ten degrees of freedom: six for rigid-body dynamics, three for rotor-flapping, and one for the rotation of the rotor (Talbot, Tingling, Decker, and Chen, 1982). The simulator was modified at the Georgia Institute of Technology to specifically model the UH-1H, and to provide a simple interface to the control software and search algorithm (a genetic algorithm) used in this research (Mittal and Prassad, 1994; Patterson, 1994).

The state variables of an aircraft are generally defined about a right-handed aircraft body coordinate system with the x-axis pointing out the nose, the y-axis pointing out the right side, and the z-axis pointing out the bottom. From this convention, the following state variables are defined:

u = dx/dt (m/sec)	Forward velocity
v = dy/dt (m/sec)	Rightward velocity
w = dz/dt (m/sec)	Downward velocity
du/dt (m/sec^2)	Forward acceleration
dv/dt (m/sec^2)	Rightward acceleration
dw/dt (m/sec^2)	Downward acceleration
p (deg/sec)	Angular velocity around the x axis
q (deg/sec)	Angular velocity around the y axis
r (deg/sec)	Angular velocity around the z axis
R_u (m/sec)	Reference (or desired) forward velocity
R_v (m/sec)	Reference rightward velocity
R_w (m/sec)	Reference downward velocity
R_Ψ (deg)	Reference heading

Aside from the variables defined in body coordinates, the yaw, pitch and roll angles are defined as Euler angles in a ground reference frame. Obviously altitude is also defined with respect to the ground reference frame. Notice that all distances are measured in meters, all angles in degrees, and time in seconds. Therefore, aside from the variables introduced above, the following variables are important in this particular problem

Ψ (deg)	Yaw (heading)
Θ (deg)	Pitch (angle from horizon to the nose)
Φ (deg)	Roll (angle from horizon to the right wing)
h (m)	Altitude

17.4 FUZZY LOGIC CONTROLLER ARCHITECTURE

The goal of the effort described here is, first and foremost, to develop a control system for a helicopter. But, above and beyond that, the goal is to develop a system that is aggressively effective, computationally efficient, and easily adaptable to new rotorcraft. Furthermore, it is desirable to have the system perform as well as a human pilot. To achieve these goals, the controller must be capable of quickly and smoothly moving the helicopter from one steady state condition to another. It must also be able to perform a simple set of maneuvers such as acceleration, deceleration, and steady turn. As with any computational system, the simpler the algorithm that accomplishes the specified tasks, the better.

Fuzzy control directly addresses many of the concerns associated with the above stated goals. In the end analysis, it is the specific architecture of the controller that determines the success or failure of the controller in achieving the goals set forth. The controller architecture will also determine the way state information flows through the controller. In the systems we have dealt with to this point in the book, the state variables were always related to the extent that we could not manipulate any information independently. However, it is important to note that information that is unrelated, or not greatly coupled, can be sent to separate processing blocks, termed "fuzzy blocks." Using such an approach accomplishes at least two things. First, this approach allows a controller to be developed that has specialized sections. Each section deals with a unique portion of the state space and addresses a specific aspect of control. Second, this approach substantially reduces the number of rules required by the fuzzy controller.

A classical fuzzy controller would typically consists of one fuzzy block with an input for each state variable (x_i) and an output for each control action (y_j) as shown in Figure 17.3a. In this architecture, each input variable is explicitly related to the other input variables through rules in the rule base. This classical fuzzy control architecture is the one we have developed in all fuzzy systems descried to this point in the book. For large problems, however, the classical approach to the development of fuzzy systems becomes impractical because the number of rules increases multiplicatively with the number of input variables. If each of the 17 variables used in the helicopter controller were described with but three membership functions, then the rule base would contain 3^{17} (approximately 129 x 10^6) rules. A fuzzy controller with a rule set of this size would be terribly slow.

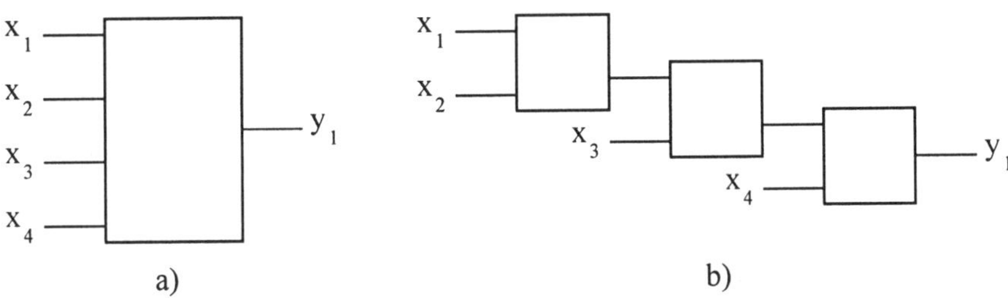

Figure 17.3 - A classical fuzzy controller (a) can often be replaced by a modular architecture (b) that performs adequately.

There is an alternative to the classical single block fuzzy control system. Consider the modular architecture shown in Figure 17.3b. Notice that the condition variables are not all handled simultaneously within a fuzzy block. Rather, the condition variables are manipulated in series based on domain knowledge indicating which input variables are most closely related. Using knowledge of the dynamics of the system to divide a process into separate control tasks allows adequate control with much fewer rules than required by the classical fuzzy controller architecture. This is a major contribution of the current chapter. If each input is described with three membership functions, the distributed controller in Figure 17.3b requires only $3^2 + 3^2 + 3^2 = 27$ rules. If the inputs are not tightly coupled, the modular controller can achieve acceptable performance with a mere fraction of the number of rules required by a more traditional fuzzy control architecture.

Developing a simplified yet suitable architecture is often quite difficult. While it is desirable to generate a controller with a small number of rules, extreme care must be taken to preserve important relationships between the state variables. A careful balance between reducing the number of rules and achieving an accurate representation of the coupling in the system, accomplished through the careful development of modular units or fuzzy blocks, is needed to create an efficient and effective controller.

Designing the helicopter controller architecture is largely a process of trial and error tempered with a minimal amount of human expertise. After several design iterations, the configuration shown in Figure 17.4a through Figure 17.4d was adopted. The controller is composed of four sections, one for each control apparatus present in the aircraft. The four sections represent control of the longitudinal cyclic (δ_{lon}) (Figure 17.4a), lateral cyclic (δ_{lat}) (Figure 17.4b), pedals (δ_{tr}) (Figure 17.4c), and collective (δ_{col}) (Figure 17.4d). Note that in these figures the extension "_dot" represents a derivative with respect to time. Notice in the figure(s) that each control goal is managed through the interaction of several fuzzy blocks.

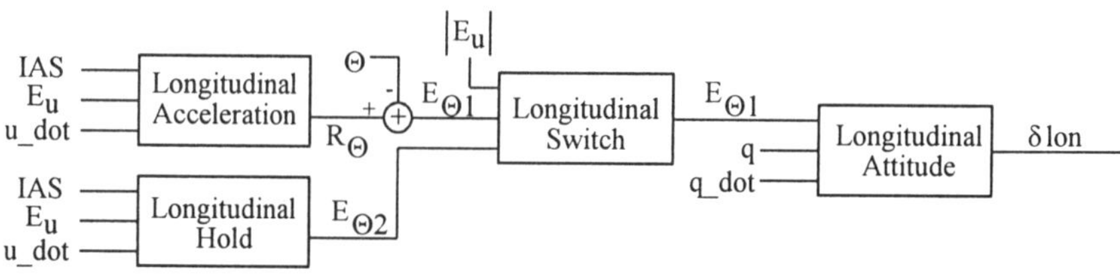

Figure 17.4a - Section of the fuzzy controller used for longitudinal cyclic control.

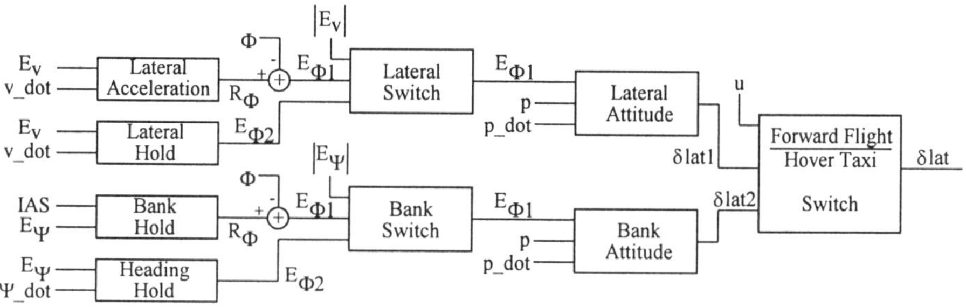

Figure 17.4b - Section of fuzzy controller used for lateral cyclic control.

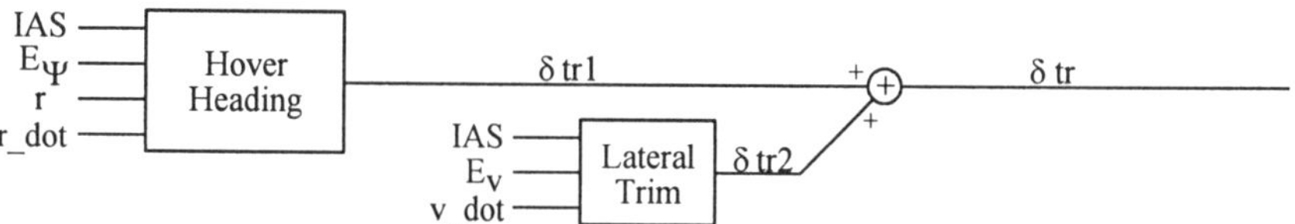

Figure 17.4c - Section of the fuzzy controller used for pedal control.

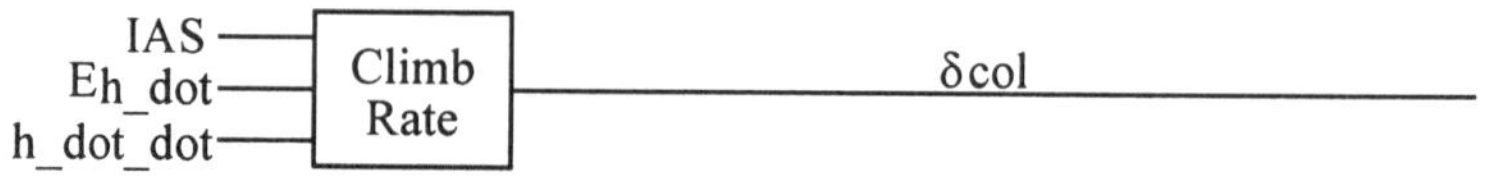

Figure 17.4d - Section of fuzzy controller used for collective control.

Although the architecture described in the above figure(s) may seem imposing, if not intimidating, the resulting system is far simpler than a control system containing the number of rules that would be required in a traditional fuzzy architecture. In the above system, the longitudinal cyclic control motion is governed by the fuzzy block labeled Longitudinal Attitude. Given the error in pitch, q, and q_dot, the Longitudinal Attitude block infers a change in longitudinal cyclic. The

Longitudinal Switch determines which of two control strategies is employed based on the error in forward velocity. If the forward velocity of the helicopter is far from the goal velocity, then the error in pitch is determined by the Longitudinal Acceleration block. Otherwise, the error in pitch is determined by the Longitudinal Hold block. Notice that the Longitudinal Acceleration block infers a desired pitch angle while the Longitudinal Hold block infers an error in pitch angle. While at first this arrangement might seem overly complex it is designed to model the decision process of a pilot. When a pilot is near the desired forward velocity, he or she achieves the velocity goal by making small adjustments to the pitch of the aircraft. When a pilot is far from the desired forward velocity, he or she will typically maintain a desired pitch attitude to accelerate or decelerate the helicopter. In both cases, the result is a function of the current airspeed, the error in forward velocity and the forward acceleration.

The task of achieving lateral cyclic control is split into two control branches (Figure 17.4b), both of which resemble the longitudinal section of the controller. The two branches represent the different functions of lateral cyclic at hover and at forward flight. At hover, the lateral cyclic is used to determine lateral velocity of the aircraft. In forward flight, however, the desired lateral velocity is almost always zero. In this case, the lateral cyclic is used to roll the aircraft into turns to achieve new headings. Notice in Figure 17.4b that the top branch of the lateral controller uses lateral velocities to achieve control at hover. The bottom branch of the lateral controller uses heading information to achieve heading control in forward flight.

Control of the tail rotor (Figure 17.4c) is also a function of the current forward airspeed. While it is not obvious from the figure, the rules in the Hover Heading block become ineffective as airspeed increases. The opposite is true for the Lateral Trim block. At hover, the Hover Heading block controls the tail rotor. In forward flight, the Lateral Trim block controls the tail rotor. Between hover and forward flight, the actions of the two control blocks are linearly combined. The purpose of the Lateral Trim block is to make very small adjustments to the tail rotor in forward flight to eliminate lateral motion, or "side-slip."

Finally, collective control (Figure 17.4d) is a function of the current airspeed, error in climb rate, and climb acceleration of the helicopter. The collective control section is the simplest yet most effective section of the controller.

Fuzzy blocks or switches are used throughout the controller to provide specialized control for different flight regimes. An example is the Forward Flight/Hover Taxi switch in the lateral cyclic controller (Figure 17.4b). If the forward velocity of the helicopter is less than 6 m/s, then the result from the top section of the lateral controller is passed through the switch. If the forward velocity is greater than 12 m/s, the result of the bottom section of the lateral controller is passed through the switch. When forward velocity is between 6 m/s and 12 m/s, a linear combination of the results of the top and bottom sections of the lateral controller are provided as output from the switch. The fuzzy rules in the switches are predetermined, not manipulated by the search algorithm during rule discovery.

There are 12 fuzzy blocks in the helicopter controller, and numerous fuzzy switches. Each control block receives several input variables, which are described by a collection of triangular membership functions. Also, the inputs are "clipped" to insure that they fall between a minimum and maximum value. These details are listed for each fuzzy control block in Table 17.1. The units for the values in the **Min** and **Max** columns of the table correspond to the units of the input variable they describe. The fuzzy switches are not listed; each switch has three inputs. The first input is the criteria for the switch, and is described by two membership functions. The other two inputs are the values to be switched between.

Table 1. Descriptions for the inputs of each fuzzy control block.

Block	Input	Number of Members	Min	Max
Longitudinal Acceleration	IAS	3	6	30
	Eu	3	-30	30
	u_dot	3	-1	1
Longitudinal Hold	IAS	3	6	30
	Eu	3	-6	6
	u_dot	3	-1	1
Longitudinal Attitude	E Theta	5	-6	6
	q	3	-10	10
	q_dot	3	-10	10
Lateral Acceleration	Ev	3	-6	6
	v_dot	3	-1	1
Lateral Hold	Ev	5	-6	6
	v_dot	3	-1	1
Lateral Attitude	E Phi	5	-6	6
	p	3	-10	10
	p_dot	3	-10	10
Bank Hold	IAS	2	30	60
	E Psi	3	-20	20
Heading Hold	E Psi	5	-10	10
	Psi_dot	3	-3	3
Bank Attitude	E Phi	7	-10	10
	p	3	-10	10
	p_dot	3	-10	10
Hover Heading	IAS	2	6	12
	E Psi	7	-30	30
	r	3	-60	60
	r_dot	3	-30	30
Lateral Trim	IAS	2	6	12
	Ev	3	-6	6
	v_dot	3	-1	1
Climb Rate	IAS	2	6	12
	Ez_dot	3	-3	3
	z_dot_dot	3	-3	3

17.5 SEARCH IMPLEMENTATION

The genetic algorithm used to determine the rule base for the helicopter fuzzy controller is the simple genetic algorithm described in Chapter 5. It employs the three classic genetic operators of *reproduction*, *mutation*, and *crossover*. To use the genetic algorithm as a search technique, two major issues must be addressed: (1) the coding of the parameters, and (2) the development of a fitness function.

Per normal, when we employ a genetic algorithm to tune, or in this case design, a fuzzy controller, we can search for the membership functions, for the rules, or for both. Here, the decision was made to search for the fuzzy rules because their values are far less intuitive than the fuzzy membership functions. These rules are encoded using a conventional multi-parameter, mapped, Gray coding (Gray, 1953), and appear in the bit string in the order they appear in the controller. The rules for each fuzzy control block are mapped using the range and number of bits shown in Table 17.2. The units for the values in the **Min** and **Max** columns of the table correspond to the units of the output variable associated with the rules they describe. Since there are 414 rules in the fuzzy controller, each mapped with 4 bits, the resulting encoded bit string is 4*414 = 1656 bits in length. Although the fuzzy architecture we are employing dramatically reduces the number of rules in the fuzzy controller, the search space of the genetic algorithm still contains $2^{1656} = 3.2 * 10^{498}$ possible solutions. If we had employed a classical fuzzy architecture, searching for the rules required would simply be impractical.

Table 17.2 - Parameters used to map the fuzzy rules into a bit string.

Block	Number of Rules	Min	Max	Bits
Longitudinal Acceleration	27	-20	20	4
Longitudinal Hold	27	-6	6	4
Longitudinal Attitude	45	-2	2	4
Lateral Acceleration	9	-6	6	4
Lateral Hold	15	-6	6	4
Lateral Attitude	45	-2	2	4
Bank Hold	6	-20	20	4
Heading Hold	15	-10	10	4
Bank Attitude	63	-2	2	4
Hover Heading	126	-2	2	4
Lateral Trim	18	-0.2	0.2	4
Climb Rate	18	-2	2	4

In most problems involving a genetic algorithm search, there is a performance versus time issue involved; performance generally improves as the number of possibilities considered increases – a move that obviously increases the time required to complete the search. The complexity (computational intensity) involved with the fitness function is the major factor in the time required to compute a solution. Here, the fitness function is composed of a weighted sum of differences. The difference between a set of states and their desired values is calculated as error at each time step. The absolute values of the errors are summed over the course of the simulation yielding a cumulative error for each of the states. These cumulative errors are combined in a weighted sum to produce the fitness. In the helicopter problem, the fitness value typically consists of a weighted sum of integrations of state error over several simulations. Several simulations with different initial conditions and goals were needed to adequately cover the parameter space being searched on see discussion in Chapter 3 on the cart-pole controller). A typical fitness function for N separate simulations lasting T seconds is defined as:

$$Fitness = \sum_{I=1}^{N}\sum_{t=1}^{T}(E_u + E_v + E_w + E_\psi) \qquad (17.1)$$

where E_u is the error in the forward velocity, E_v is the error in the rightward velocity, E_w is the error in the downward velocity, and E_Ψ is the error in the reference heading.

Since the fitness function in this problem is so computationally intensive, the genetic algorithm was applied in distinct stages to search for the fuzzy rules. In the first stage, the genetic algorithm was employed to search the entire parameter space for a solution that was not immediately divergent. In this stage, the only region of flight considered was hover. In the second stage, each section of the controller was trained independently. First, all rules except those in the Longitudinal Attitude block were excluded from the search. The Longitudinal Attitude block was trained using a fitness function that evaluated short simulations from several sets of initial conditions to various attitude goals. In this way, the control block was trained to reach and maintain a desired pitch angle. Next, the rules for the Longitudinal Hold block were discovered. The fitness function consisted of several simulations at different airspeeds. The criteria for evaluating quality performance was based on how well the controller could maintain airspeed despite perturbations.

At this point, the controller could maintain forward airspeed. The next step was to train the Longitudinal Acceleration block to allow airspeed changes. Once again, the fitness function entailed several simulations in which the controller was required to accelerate and decelerate the helicopter through a range of airspeeds. This procedure was repeated for the other three sections of the controller. A few iterations of this process were required to achieve good performance throughout the flight envelope.

It is important to note that the training of the fuzzy controller was not accomplished on a personal computer. To solve this problem in a timely manner, the genetic algorithm was implemented on a network of approximately 40 workstations. Even then, the design of the controller took on the order of a week to complete.

17.6 RESULTS

The helicopter fuzzy logic controller was developed to perform several maneuvers throughout the flight envelope of the UH-1H helicopter. Effective rules for the fuzzy controller were discovered using a genetic algorithm and a numerical model of a UH-1H helicopter. The performance of the controller was evaluated both in simulation and in actual flight tests conducted at White Sands Missile Base in conjunction with the U.S. Army.

Its successful performance in controlling the model indicates the effectiveness of a genetic algorithm in generating fuzzy logic rules. Figure 17.5 shows the results of the controller performing a simulated hover. In this figure, u, v, and w are the forward, rightward, and downward velocities of the helicopter measured in meters per second, respectively. The heading (Ψ) is measured in degrees. After a brief period of transition, the controller brings all velocities and the heading to zero.

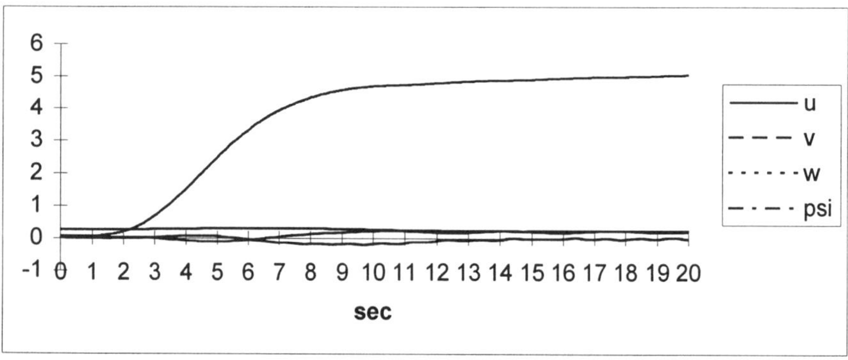

Figure 17.5 - The fuzzy controller is able to effectively achieve hover in simulation.

Figure 17.6 shows data for a transition from hover to 5 m/s forward taxi. The heading and lateral and vertical velocities are held near zero. In forward flight, a coordinated turn is performed by flying the helicopter to a new bearing without allowing lateral velocity. Figure 17.7 shows data for a coordinated turn at 34 m/s forward airspeed. The airspeed is maintained through a right turn of 60 degrees.

Climb rate is held at zero for the duration of the turn. There is, however, oscillation in the lateral velocity. This is a result of a flaw in the controller design. The pedals are used to control heading at hover, but need not be used in forward flight except to trim the aircraft during transition to forward flight. The Lateral Trim control block in the helicopter fuzzy controller constantly tries to correct lateral velocity errors in forward flight using the pedals. The fuzzy controller should be redesigned to provide trim during transition only.

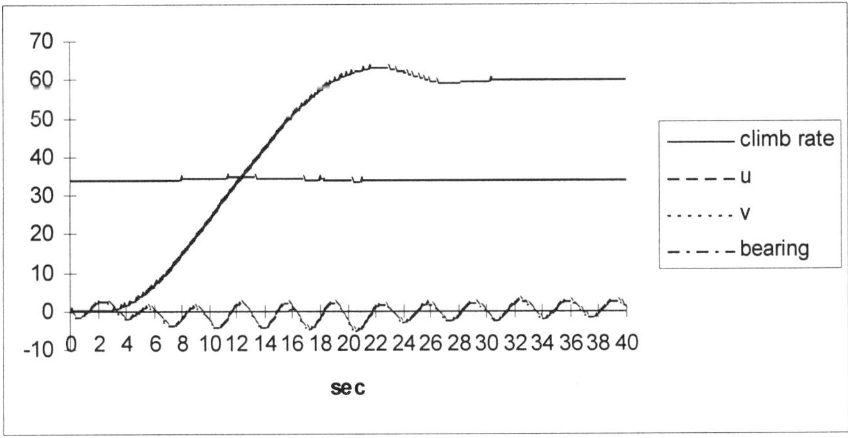

Figure 17.6 - The fuzzy controller achieves the goal of forward taxi to 5 m/s in simulation.

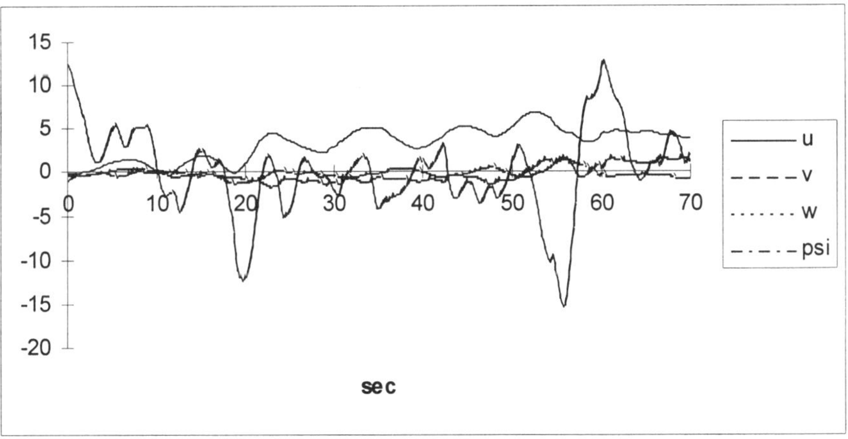

Figure 17.7 - The fuzzy controller successfully completes a coordinated turn in simulation.

Since the fuzzy logic controller performed well in simulation, it was used in an actual flight test. Because there are differences between the dynamics represented by the numerical model and the dynamics of the actual aircraft, the successful performance of the fuzzy controller on the aircraft in actual flight tests indicates the robustness of fuzzy logic used for flight control. Figure 17.8 shows the controller performing a hover in actual flight. It successfully holds all velocities and heading about zero for 90 seconds.

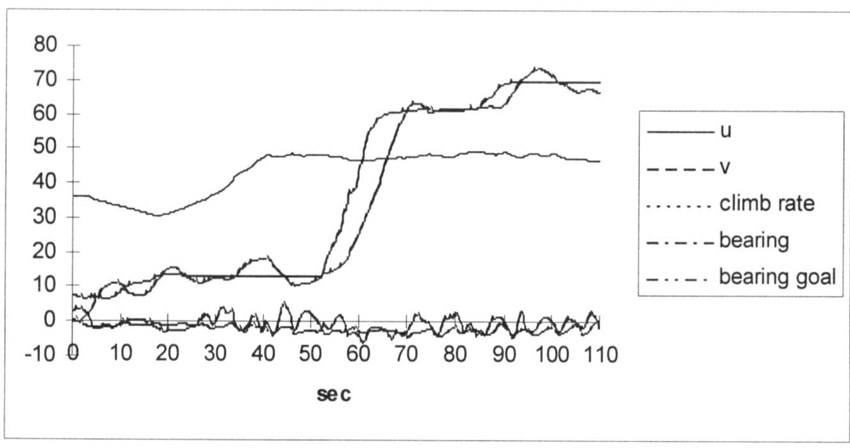

Figure 17.8 - The fuzzy controller achieves hover in actual flight.

17.7 REVIEW AND PREVIEW

A fuzzy system capable of controlling a helicopter effectively and aggressively was successfully developed using a genetic algorithm to discover the fuzzy rules. Quality performance was demonstrated in both simulations and actual flight tests. The fuzzy controller architecture developed is general enough to be applicable to a variety of rotorcraft. Moving the controller to a new helicopter simply requires discovering appropriate fuzzy rules with the genetic algorithm.

This chapter has addressed one of the issues that can occur in the development of real-world fuzzy systems: how does one handle large numbers of condition variables yet keep the rule set from becoming too large? In the helicopter flight control application addressed here, we have proposed exploiting domain knowledge to account for coupling effects between various variables. In this way, we have been able to limit the size of the rule base substantially. Unfortunately, this approach means that each problem is slightly different, but not so different that it cannot be used for different helicopters. However, when the rubber meets the road, isn't this always the situation?

Thus far in this book we have viewed all of the problems of designing and adapting fuzzy controllers as straight optimization problems. However, there is an alternative approach that has merit and warrants mention. Thus, in the next chapter we will describe a system that employs a genetic algorithm to accomplish machine learning, a learning classifier system. Learning classifier systems evolve an entire rule set from scratch, neglecting completely rules that are unnecessary. We will describe an approach whereby learning classifier systems can be made to manipulate fuzzy production rules rather than their usual crisp rules. We will apply this alternative approach to fuzzy rule-based design to a liquid level system.

REFERENCES

Gray, F. (1953, March 17). *Pulse code communication*, U. S. Patent 2,632,058.

Mittal, M., & Prassad, J. V. R. (1994). *Modeling the UH-1H helicopter using the ARMCOP program*. NASA Contract L27548DA.

Saunders, G. H. (1975). *Dynamics of helicopter flight*. New York: John Wiley & Sons.

Sugeno, M., Griffin, M. F., & Bastian, A. (1993). Fuzzy hierarchical control of an unmanned helicopter. *Proceedings of the IFSA Congress*, Korea

Talbot, P., Tingling, B., Decker, W., & Chen, R. (1982). *A mathematical model of a single main rotor helicopter for piloted simulation*. NASA TM 84281.

Chapter 18

Fuzzy Classifier System

18.1 INTRODUCTION

The major thread throughout this book has been the idea that genetic algorithms can be used to design and adapt fuzzy control systems. We have investigated other ideas such as the accuracy of fuzzy systems, the effect of using various membership function forms, and the steps that can be taken to prevent fuzzy rule sets from growing too large. However, even in addressing these issues our default position has been to employ a genetic algorithm as an optimization routine to design a fuzzy controller to meet some predefined standards. In this chapter we will see that genetic algorithms can be used in a far different way to develop fuzzy systems.

This chapter discusses the development of a fuzzy classifier system that discovers rules for controlling a mathematical model of a pH titration system. Fuzzy classifier systems successfully combine the strengths of learning classifier systems and fuzzy controllers. Learning classifier systems resemble familiar production rule-based systems, but they represent their IF-THEN rules by strings of characters rather than in the traditional linguistic terms. Like fuzzy controllers, fuzzy classifier systems encapsulate knowledge in the form of production rules. The results presented in this chapter demonstrate the ability of fuzzy classifier systems to generate a fuzzy control system.

A fuzzy classifier system was first proposed by Valenzuela-Rendón (1991) to map functions of continuous variables. This chapter describes an extension to the work of Valenzuela-Rendón in that the fuzzy classifier system is used to solve a process control problem. The fuzzy classifier system described represents a method for achieving truly intelligent computer systems that are capable of manipulating complex problem environments because the fuzzy classifier system, in effect, learns to efficiently manipulate a pH titration problem environment without prior knowledge of an appropriate set of rules. This is different from the approach presented up to this point because the fuzzy systems of previous chapters required some information about the rules.

Fuzzy classifier systems consist of three main components: (1) a rule and message system, (2) an apportionment of credit system, and (3) a genetic algorithm. The rule and message system is a mechanism by which the fuzzy classifier system interacts with the problem environment. The fuzzy classifier system receives information concerning the condition of the problem environment and takes an action on the environment based on its rule set. A mechanism exists in the fuzzy classifier system for gauging the current state of the system to be controlled, and for effecting actions on the system.

In the apportionment of credit system, rules compete for the right to take their action on the problem environment, and are rewarded or punished in accordance with their performance. At first glance, this system seems to be rather straightforward. The apportionment of credit system assigns a relative merit to each rule based on the rule's effectiveness, and this value is much like the conventional fitness function of traditional genetic algorithms (it is this value of merit that is to be maximized). However, difficulties occur when rules prescribe actions that do not produce immediate change in the system. The apportionment of credit system must be able to give credit to rules that caused changes in the present, as a result of past actions.

In this chapter, the fuzzy classifier system is used to control a specific problem environment: a simple pH titration system (simpler than the pH system used earlier in this book). The fuzzy classifier system developed to control a computer simulation of the pH system is quite effective in achieving the process control objective. Its performance is compared to that of a fuzzy controller that has been shown to manipulate the pH environment in an effective manner. The performance of the fuzzy classifier system demonstrates the potential of this approach to adaptive process control. Its effectiveness in the highly nonlinear problem of pH control points to far-reaching implications for application in diverse industrial fields such as mineral processing, chemical engineering, and solid waste disposal.

18.2 THE PHYSICAL SYSTEM

A fuzzy classifier system is used to control a simple laboratory pH system. A schematic of this pH system is shown in Figure 18.1. The system is basically the core component of the pH system considered in Chapter 14. It consists of a tank initially containing a given volume of a solution having a known pH. There are two valved input streams into the beaker. The valves on these two ***control input streams***, one a strong acid (0.1 M HCl) and one a strong base (0.1 M NaOH), can be adjusted to cause a change in the pH of the solution in the tank. The objective of the control problem is to neutralize the solution — drive the pH to 7 — in the shortest time possible by adjusting the valves on the control input streams. Additionally, the valves on the control input streams are to be fully closed after the solution is neutralized. As a constraint on the control problem, the valves can only be adjusted a limited amount (0.5 mL/s/s, which is 20 pct of the maximum flow rate of 2.5 mL/s), to restrict pressure transients in the associated pumping systems. The computer model of Chapter 14 is applicable to the system of Figure 18.1.

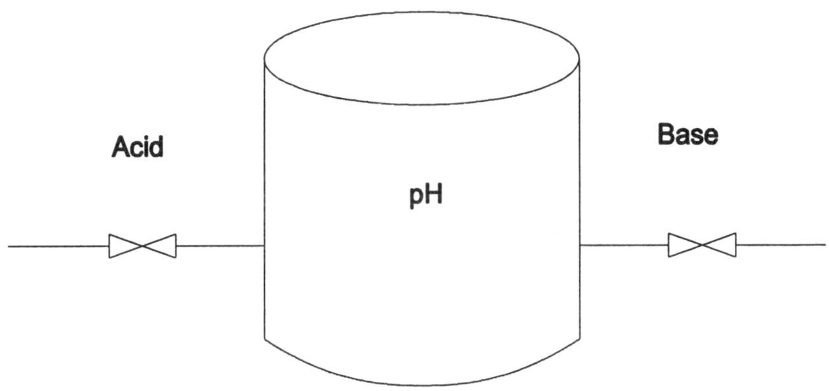

Figure 18.1 - The above pH titration system is highly nonlinear and provides a challenging problem environment on which to test the effectiveness of the fuzzy classifier system.

18.3 LEARNING CLASSIFIER SYSTEMS

A classifier system is a genetics-based machine learning system that learns rules, called classifiers, to govern its performance in a given environment e.g., the pH titration system. The systems are based on a model of a service economy in which money is exchanged for services, and include a mechanism for discovering new rules. Unlike in traditional expert systems, a rule must earn the right to be implemented, not fixed by the programmer. In classifier systems the rules are forced to coexist in a service economy where a competition is held to decide which rule will be put into effect under a specified set of conditions in the environment. The competitive nature of the economy ensures good rules survive and bad rules die off, while the exploratory portion of the learning classifier system creates new rules. These systems operate incrementally, testing new rules while steadily improving performance in an environment.

In general, classifier systems are composed of three subsystems:

1) Rule and message system;
2) Credit assignment system;
3) Rule discovery system (genetic algorithm).

These three subsystems, when combined with a mathematical model of the environment, form a computationally complete system capable of learning effective rules for interacting in an environment and controlling processes. The mathematical model is used to investigate the effectiveness of untested rules.

The rule and message system is similar to a production system. Production systems are schemes that use rules as their only means of operation. The rules are generally of the form:

IF {*condition*} **THEN** {*action*}.

When the ***condition*** exists in the environment, then the ***action*** is taken. In learning classifier systems the rule and message system must complete the following basic tasks. The learning classifier system evaluates the existing state of the environment; it determines what conditions exist. These conditions are compared with the current rule set to determine which rules are eligible to be put into effect. A competition is held among the eligible rules and a set of winning rules is selected. These rules take their associated action by sending a message to the problem environment thereby causing a change in the environment. It should be noted that the first competition is strictly a random decision since all rules initially have the same relative strength. However, as will be seen shortly, all future competitions are decided based upon the previous successes or failures of individual rules. Once the outcome of this competition has been determined, the rule and message system becomes temporarily inactive and the credit assignment system takes over. Before the credit assignment system is discussed, note that the rule and message system is designed to activate several classifiers in parallel, which means that a learning classifier systems is a parallel production system even though it is generally implemented on sequential machines.

The purpose of the credit assignment system (Minsky, 1967) is to evaluate how useful each classifier is in producing desirable effects on the environment (how useful it is in solving the problem at hand). The method used most often is the bucket brigade algorithm of Holland (1962). In the bucket brigade algorithm every classifier is assigned a strength that represents that classifier's usefulness in solving the problem. The classifiers that are eligible to take an action at a given time step bid a portion of their associated strength for the right to take their action. Once all the eligible rules have made their bids, winners are selected based on the size of the bids; those that bid highest have the highest chance of being selected to take their action. If the environment is in a more desirable state than at the previous time step, the auction winners pay their bids to the classifiers that took action at the previous time step. Thus, the rules that caused desirable changes in the environment are rewarded with payoff. A mandatory payment is also received from every classifier at each time step. This payment, or tax, helps weed out poor rules because in the rule discovery system a classifier's existence depends on its associated strength. If a classifier fails to win the auction and receive a reward, its chance of survival is reduced. The bucket brigade algorithm's redistribution of classifier strengths leads to bids that are representative of classifiers' potential for achieving the goal; effective rules bid proportionately higher.

The purpose of the rule discovery system is to generate new rules with potential for more efficiently reaching the goal assigned to the learning classifier system. The genetic algorithm is the technique generally used in rule discovery systems. The genetic algorithm requires the elements of the search space (in this case the rules) to be coded as finite length strings. Effective classifiers, those with high associated strengths, are selected for combination with other highly fit classifiers. Portions of the classifiers are selected at random and combined with portions of other classifiers through the standard genetic algorithm operators of reproduction, crossover, and mutation to produce new rules. The intent is to combine advantageous qualities from two separate classifiers to form two new, more highly fit classifiers.

The three subsystems discussed in this section form a system capable of learning and evaluating new rules for interacting in an environment. A more detailed description of learning classifier systems can be found in Goldberg (1989).

18.4 APPLICATION OF A FUZZY CLASSIFIER SYSTEM

The first step in producing a fuzzy classifier system for the pH titration system is to develop a means of representing linguistic rules and fuzzy membership functions as strings of characters. For this study a direct mapping was used to represent the rule sets. The condition portion of each classifier is a function of the current pH and the current time rate of change of pH, ΔpH. The action portion of each classifier adjusts the net inflows (Q_{ACID} and Q_{BASE}) at the following time step. The coding used for the rule set is summarized in Table 18.1. It is important to note that Table 18.1 includes only the coding used for the rules. The character strings used contained information concerning both the rules and the membership functions and were 69 bits long.

Table 18.1. - The coding used to represent the rule sets.

Condition				Action			
pH		ΔpH		Q_{ACID}		Q_{BASE}	
binary	set	binary	set	Binary	Set	binary	set
000	Very Acidic	00	NL	00	Z	00	Z
001	Acidic	01	NS	01	S	01	S
010	Marginally Acidic	10	PS	10	M	10	M
011	Neutral Acidic	11	PL	11	L	11	L
100	Neutral Basic						
101	Marginal Basic						
110	Basic						
111	Very Basic						

The basic binary coding in which each bit position is characterized by either a 1 or a 0 would be sufficient for the learning classifier system. However, to add flexibility and allow for more general rule representation, a third character is introduced, namely the # or "don't care" symbol. A bit position filled by a # is matched by either a 1 or a 0. An example rule follows:

condition --> action
00#00 --> 0011.

In this rule, the value of pH at the current time is defined by the substring 00#. This condition is satisfied by either 000 or 001. Thus, this rule says that if the pH is currently very acidic or acidic with a negative large time rate of change of pH, then minimize the flow of acid into the tank and maximize the flow of base into the tank at the next time step.

Now that a means of representation of the rules and fuzzy membership functions has been established, the implementation of a fuzzy classifier system is discussed. A rule set of 250 rules was used to control a mathematical model of the pH titration system. These rules are used to manipulate the computer model of the pH titration system until 25,000 rules have been evaluated. At the end of this period, 120 new rules are produced using the reproduction, crossover, and mutation operators of a genetic algorithm. These 120 rules replace 120 of the 200 poorest rules in the previous rule set. By crowding out the poor rules and relying on the genetic algorithm' survival-of-the-fittest approach, the rule sets achieve higher levels of performance. A learning cycle consisted of the 25,000 evaluations or one generation of new rules using the genetic algorithm. However, learning actually takes place after each individual rule activation, at which time the bucket brigade algorithm redistributes payment to the rules. The steps comprising these learning cycles are relatively straightforward and consist of the following:

1) Evaluate the condition or state of the physical system at the current time step.

2) Compare the state of the physical system to the condition portion of each of the 250 classifiers to determine which rules are eligible to take their action.

3) Hold an auction between the eligible rules to determine which rules get to take their actions. The bid placed by a given rule is defined by

$$BID = S * C_{bid}$$

where BID is the rule's bid, S is the rule's strength, and C bid is a constant defined in this instance to be 0.1. Within the model of a service economy, strength becomes a measure of how effective a rule is at driving the system to its setpoint.

4) Evaluate the effectiveness of the action. If the action, which is a function of both the rules and the fuzzy membership functions, drives the system closer to its setpoint, then the rules that won the auction at the previous time step are rewarded with the current winners' bids and a tax paid by each classifier as described below. Otherwise no rule receives a reward and the tax is accumulated in a fund and carried over to the next time step.

5) Tax each of the rules in the current rule set. Every rule is taxed according to the following:

$$TAX = (C_{bidtax} - C_{lifetax} S)$$

where TAX is the tax paid by a classifier, C_{bidtax} is a constant equal to $5*10^{-5}$ if the rule's condition is met and thus it placed a bid (it is equal to 0.0 otherwise), and $C_{lifetax}$ is a constant equal to $2*10^{-5}$ under every circumstance.

6) After 25,000 actions are taken, apply the genetic algorithm to generate a new rule set that includes 120 new rules. The genetic algorithm combines portions of the most effective rules as determined by their strength, which rises if a rule is effective and falls if a rule is non-effective due to the nature of the apportionment of credit algorithm.

Although the mechanics are relatively simple, the fuzzy classifier system is able to employ the exploratory power of the genetic algorithm and the structured credit assignment of the bucket brigade to learn new and improved rules for controlling pH titration system.

18.6 RESULTS

Evaluating the effectiveness of a process control system is not a trivial endeavor; there are various criteria for efficient control that can be established, and there are numerous conditions for which the controller must be able to perform. In evaluating the performance of the fuzzy classifier system it is important to keep in mind the control objective, which is to neutralize the solution as fast as possible while not violating the constraints placed on the flow rates. Additionally, it is important to realize that if the controller can accomplish the control goal from extreme portions of the control space (when the solution is initially extremely acidic or extremely basic), it should perform well in the more moderate portions of the control space. The performance of the fuzzy classifier system is compared to the performance of a fuzzy controller that was developed by the authors for the pH titration system.

Figure 18.2 summarizes the performance of the fuzzy classifier system. In the particular case depicted, the fuzzy classifier system is posed with the problem of neutralizing a basic solution by acid addition. As can be seen in the figure, the fuzzy classifier system rapidly locates rules for driving the pH to values that range roughly between 5.2 and 8.5, which is the region in which a titration curve is most nonlinear. Then, after a brief period of exploration, the controller is able to evolve into a form that forces the pH to the desired value of 7. Figure 18.3 shows the performance of a fuzzy controller that was designed by the authors to solve the pH problem. Note that this controller drives the pH to a value that is between 6.5 and 7.5. It is important to note that the performance of the author-developed fuzzy controller can be improved by altering the membership functions (Karr and Gentry, 1993), but this is a time-consuming task to accomplish without the assistance of some computational algorithm such as a genetic algorithm. Nonetheless, the two figures demonstrate the fact that the fuzzy classifier system contains a mechanism for improving its performance through the discovery aspects of a genetic algorithm.

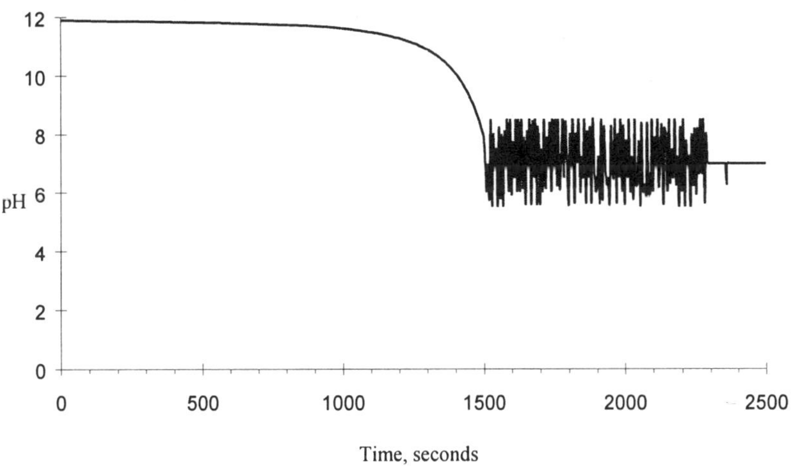

Figure 18.2 - The fuzzy classifier system neutralizes a basic solution.

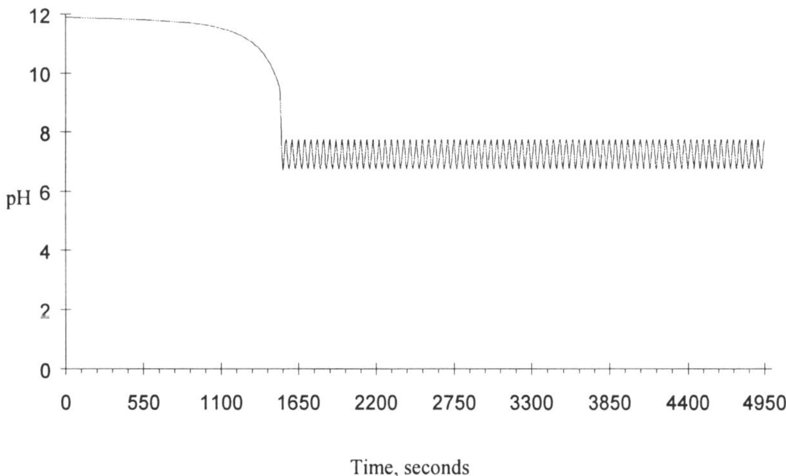

Figure 18.3 - A fuzzy controller neutralizes a basic solution.

Figures 18.2 and 18.3 demonstrate the effectiveness of the fuzzy classifier system in the pH problem environment. Although the results presented do not alleviate concerns as to the ability of the fuzzy classifier system to perform effectively in process control problems, they do point to the potential of these fuzzy rule discovery systems. These results demonstrate the ability of the system to discover both rules and fuzzy membership functions for effectively manipulating a highly nonlinear physical environment.

18.7 REVIEW AND FUTURE WORK

The previous section provided results that indicate a fuzzy classifier system is capable of efficiently controlling a pH titration system. However, there are still a number of research issues left open. The following topics are currently under investigation:

- The pH titration system is being extended to include external perturbations such as additions of a buffering solution. This extension will make the pH titration system more like chemical systems currently used in industry, and will force the fuzzy classifier system to discover rules in real time more often than it has to with the current physical system.

- A general purpose credit assignment system is being developed. Currently, the credit assignment system is specialized for the pH titration system. For the fuzzy classifier system to be easily applied to alternative process control problems, a general purpose credit assignment algorithm must be developed.

- The fuzzy control aspect of the fuzzy classifier system is being improved. The center of area algorithm for selecting a single crisp action is but one of a number of potential solution algorithms that have been proven effective in fuzzy controllers. Current efforts are centered on including fuzzy singletons into the fuzzy classifier system. Successful implementation of this algorithm should improve the effectiveness of the fuzzy actions prescribed by the rule sets.

- Alternative schemes for encoding the classifiers are being considered. The effectiveness of the genetic algorithm's search is highly dependant on the coding scheme used. Although the current coding seems to be effective, recent research efforts (Karr, 1991) indicate that there may be more efficient ways to encode the information concerning the rules and the fuzzy membership functions.

- A micro genetic algorithm is being incorporated into the fuzzy classifier system. The micro genetic algorithm is a small-population genetic algorithm that has been reported to perform well across a spectrum of search problems, and in the problem of locating fuzzy membership functions in particular (Karr and Gentry, 1993). If this effort is successful, the fuzzy classifier system should be able to locate effective rules in less time than it currently takes.

The material in this book has provided the reader with an overview of a project conducted at the U.S. Bureau of Mines in which many facets of artificial intelligence were woven together to solve industrial problems. Much of the synergy realized to accomplish the research effort is applicable to numerous other scientific problems. However, there is still much to be done. In the next chapter we provide a very brief overview and address some of aspects of this work that we believe are fertile research areas.

REFERENCES

Goldberg, D. E. (1989). *Genetic algorithms in search, optimization, and machine learning*. Reading, MA: Addison-Wesley.

Holland, J. H. (1962). Outline for a logical theory of adaptive systems. *Journal of the Association of Computing Machinery*, **3**, 297-314.

Karr, C. L. (1991). Genetic algorithms for fuzzy controllers. *AI Expert*, **6**(2), 26-33.

Karr, C. L., and Gentry, E. J. (1993). Fuzzy control of pH using genetic algorithms. *IEEE Proceedings on Fuzzy Systems*, **1**(1), 46-53.

Minsky, M. L. (1967). *Computation: Finite and infinite machines*, Englewood Cliffs, NJ: Prentice Hall.

Valenzuela-Rendón, M. (1991). The fuzzy classifier system: A classifier system for continuously varying variables. In K. Belew and L. Booker (Eds.), *Proceedings of the Fourth International Conference on Genetic Algorithms*, 346-353.

Index

M

N

P

R

S

T

V

W